Colli

Student Support Materials for AQA

A2 Geography

Unit 3

Contemporary Geograp

Banks and Paula Howell Evans
Series Editor: David Redfern

Published by Collins Education
An imprint of HarperCollins Publishers
77-85 Fulham Palace Road
Hammersmith
London
W6 8JB

Browse the complete Collins Education catalogue at
www.collinseducation.com

10 9 8 7 6 5 4 3 2 1

ISBN 978 0 00 741572 4

Philip Banks and Paula Howell Evans assert their moral rights to be identified as the authors of this work.

British Library Cataloguing in Publication Data.

A Catalogue record for this publication is available from the British Library.

Commissioning Editor: Lucy McLoughlin

Project Editor: Lucien Jenkins

Design and typesetting: Hedgehog Publishing Ltd

Cover Design: Angela English

Index: Indexing Specialists (UK) Ltd

Production: Simon Moore

Printed and bound by L.E.G.O. S.p.A., Italy

Acknowledgements

Every effort has been made to contact the holders of copyright material, but if any have been inadvertently overlooked the publishers will be pleased to make the necessary arrangements at the first opportunity.

Text:

p. 43 © Chilworth Conservation/chilworthconservation.co.uk; p. 72 © OECD/oecd.org; p. 73 © UN/mdgs.un.org; p. 85 © Crown/statistics.gov.uk.

Photos and illustrations:

Cover and p. 1 © Alberto Pomares/istockphoto.com; p. 8 © Collins Geo; p. 11 © Collins Geo; p. 14 and p. 92 © Shi Jianping/ChinaFotoPress/Getty Images/gettyimages.co.uk; p. 17 © Collins Geo; p. 21 © Collins Geo; p. 26 © Collins Geo; p. 31 © NASA/nasa.gov; p. 39 NASA images created by Jesse Allen, using Landsat data provided by the United States Geological Survey and ASTER data provided courtesy of NASA/GSFC/METI/ERSDAC/JAROS, and U.S./Japan ASTER Science Team; p. 39 © Mongabay/Mongabay.com; p. 43 © Chilworth Conservation/chilworthconservation.co.uk; p.44 © IUCN/iucn.org; p. 48 © MARKABOND/Shutterstock; p. 49 © Collins Geo; p. 53 © Mike Peel/mikepeel.net; p. 54 © Martin Bradwick/The Photolibrary Wales/Alamy; p. 55 © Mike Wheatley/btcv.org.uk; p. 57 © Martin Cogley/martincogley.com; p. 58 © Mike Peel/mikepeel.net; p. 62 © Collins Geo; p. 67 © Collins Geo; p. 71 © Collins Geo; p. 78 © Wakefield Council/www.wakefield.gov.uk; p. 80 © Collins Geo; p. 86 © Collins Geo; p. 88 © World Bank/worldbank.org; p. 90 ©World Health Organization/who.int; p.139 © Collins Geo.

Contents

Plate movement

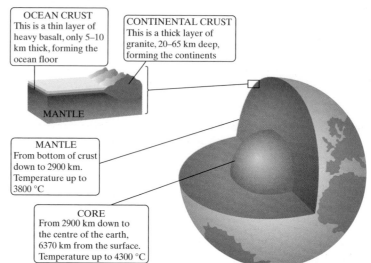

OCEAN CRUST
This is a thin layer of heavy basalt, only 5–10 km thick, forming the ocean floor

CONTINENTAL CRUST
This is a thick layer of granite, 20–65 km deep, forming the continents

MANTLE

MANTLE
From bottom of crust down to 2900 km. Temperature up to 3800 °C

CORE
From 2900 km down to the centre of the earth, 6370 km from the surface. Temperature up to 4300 °C

Fig 1
The internal structure of the earth

Essential notes

Much of the internal structure of the earth was discovered by studying the way earthquake waves are bent as they pass through the earth.

Continental crust is made of older and less dense rocks such as granite. The most abundant minerals in continental crust are aluminium and silica. The continental crust is generally 35–70 km thick and mostly over 1500 million years old.

Oceanic crust is made of younger rocks like the ones found along the Mid-Atlantic Ridge. The most common rocks are basaltic, which are denser than the ones found in the continental crust. The oceanic crust is only about 6–10 km thick and has an average age of 200 million years.

Plate tectonics theory and continental drift

This includes:
- Fossil evidence: the mesosaurus (one of the first marine reptiles) was found in both South America and South Africa
- Accurate measurement of the ocean floors
- Earthquake evidence indicated that:
  – There were two types of crust – oceanic and continental
  – There was a low velocity zone or **asthenosphere** just below the crust. This showed that it was plastic and could flow
  – Earthquake foci of differing depths followed clearly defined active zones

Essential notes

Plate tectonics theory emerged from the early work done by Wegener.

In 1912 Alfred Wegener, a German meteorologist, came up with the idea of **continental drift**. He was able to produce evidence for his theory but was derided by the scientific establishment because he could offer no mechanism that was able to move continents. During the Second World War developments in echo-sounding meant that scientists were, for the first time, able to get an accurate view of the bottom of the oceans. There were discoveries of trenches, mid-ocean ridges, seamounts, etc. By the 1950s Wegener's work was being studied once again and scientists came up with the idea of **seafloor spreading**. This is the mechanism whereby oceanic crust is created and also destroyed.

Evidence for seafloor spreading

Palaeomagnetism: This proved to be some of the most conclusive evidence. It was already known that from time to time the earth has reversed its magnetic field, the north and south poles swapping over. It was also known that as volcanic rocks cool, all the minute particles of iron in them record the magnetic field around them, preserving the respective positions of the north and south poles like a fossil.

Scientists discovered that the rocks not only showed bands of north and south magnetic fields, but that the patterns on either side of the Mid-Atlantic Ridge were almost identical. This showed that rocks were moving away from the middle of the ocean.

Then it was discovered that parts of the earth's **crust** were being destroyed along the edges of the Pacific Ocean. Along the Pacific coast, submarines had found deep **ocean trenches** leading to the discovery that the earth's surface was slowly moving into these trenches and vanishing as fast as new surface was being created along the ocean ridges.

This all led to the theory of plate tectonics. This suggests that the crust of the earth is split up into seven large plates and several smaller ones, all of which are able to slowly move around on the earth's surface. They float on the semi-molten **mantle** rocks.

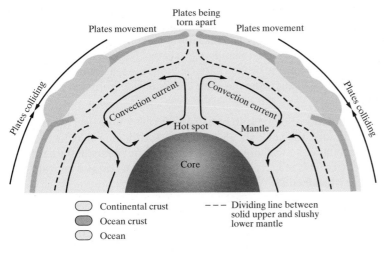

⬭ Continental crust	- - - Dividing line between solid upper and slushy lower mantle
⬭ Ocean crust	
⬭ Ocean	

The centre of the earth (the core) is very hot, and some of this heat moves outwards into the mantle in **convection currents**. As the rocks slowly move below the earth's surface, they drag the crust that lies over them. This causes the continents to move.

Essential notes

Task: Take at least three pieces of evidence (other than palaeomagnetism) from p. 4 and research them.

Essential notes

Create a list of rules that must apply if plate tectonics theory is to work, for example: *'There can never be gaps between plates, so if two plates move apart, as in the middle of the Atlantic, new rock will be formed to fill the space.'*

Fig 2
Convection currents in the mantle

Destructive, constructive and conservative plate margins

There are three types of plate margins or boundaries:

- **Constructive** (divergent) boundaries – where new crust is generated as the plates pull away from each other. These are found at mid-oceanic ridges
- **Destructive** (convergent) boundaries – where crust is destroyed as one plate dives under another
- **Conservative** (transform) boundaries – where crust is neither produced nor destroyed as the plates slide horizontally past each other

Constructive plate margins

These are located over rising convection currents in the mantle. Partially melted upper mantle rises and accumulates in a magma chamber, which may push the crust upwards causing rifts in the crust and shallow earthquakes. Eventually the magma rises through to the surface, forcing the two sides of the rift apart and creating new crust (this is the stage reached in the Great African Rift Valley). The resultant **rift valley** continues to widen and eventually fills with sea water. The process continues and the sea floor spreads to form an ocean as the plates get bigger.

The Mid-Atlantic Ridge is an example of a constructive plate margin. It is a submerged mid-ocean ridge, a mountain range running north/south for 15 000 km through the Atlantic Ocean from Iceland to a point 7 200 km east of southern South America. It breaks the ocean's surface in several places, forming groups of volcanic islands, e.g. Iceland. Most of the ridge lies 3 000–5 000 m below the ocean's surface. From the sea floor, the mountains rise 1 000–3 000 m high, and measure 1 500 km wide from east to west at their base.

The Mid-Atlantic Ridge is split by a deep rift valley along its crest, 10 km wide with walls 3 000 m high. This rift valley is widening at a rate of 3 cm per year. Where the sea floor spreads, **magma** from beneath the earth's surface rises. This magma becomes new ocean crust on and beneath the sea floor when it cools.

Destructive plate margins

These occur where two plates converge and one sinks under the other and is destroyed. The two main types of destructive plate margin are:

- Ocean/continent boundary
- Island arcs (oceanic convergence)

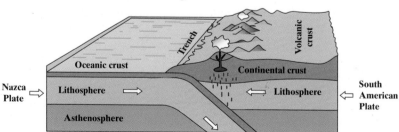

Fig 3
Oceanic-continental convergence

Ocean/continent boundary: Part of the ocean floor is dragged down with the sinking crust to form a very deep trench, parallel to the coast, e.g. the Peru/Chile trench.

Off the coast of South America along the Peru/Chile trench dense oceanic crust (the Nazca Plate) meets less dense continental crust (the South American Plate) and sinks (subducts) beneath it, dragging down the seabed to form the trench. In turn, the overriding South American Plate is being lifted up, creating a chain of **young fold mountains**, the Andes. The friction between the sinking oceanic plate and the continental plate creates great heat and melting, resulting in volcanoes along the length of the mountain chain. Strong, destructive earthquakes and the rapid uplift of mountain ranges are common in this region.

Island arcs (oceanic convergence): When two oceanic plates converge, one usually sinks (subducts) under the other and in the process a trench is formed, e.g. the Marianas Trench marks where the fast-moving Pacific Plate converges with the slower-moving Philippine Plate. The Challenger Deep, at the southern end of the Marianas Trench, is the deepest part of the ocean at nearly 11 000 m.

Essential notes

The Andes are a complex series of parallel chains of peaks (*cordillera*) with intermontane valleys. Task: research the geography of the Andes for a case study of young fold mountains.

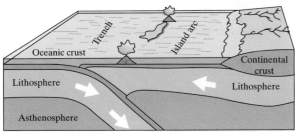

Fig 4
Island arc plate boundary

Subduction processes in oceanic-oceanic plate convergence also result in the formation of volcanoes which, over millions of years, pile up lava on the ocean floor until a submarine volcano rises above sea level to form an island volcano. Such volcanoes are typically strung out in curved chains called island arcs. Magmas that form island arcs are produced by the partial melting of the descending plate and/or the overlying oceanic lithosphere. The descending plate also provides a source of stress as the two plates interact, leading to frequent, moderate to strong, earthquakes.

Hotspots

The vast majority of earthquakes and volcanic eruptions occur near plate boundaries, but there are some exceptions. For example, the Hawaiian Islands, which are entirely of volcanic origin, have formed in the middle of the Pacific Ocean more than 3 200 km from the nearest plate boundary.

The Hawaiian Island chain has resulted from the Pacific Plate moving over a deep, stationary **hotspot** in the mantle, located beneath the present-day position of the island of Hawaii. Heat from this hotspot produces a constant source of magma by partially melting the overriding Pacific Plate. The magma, which is lighter than the surrounding solid rock, then rises through the mantle and crust (thermal **plume**) to erupt onto the sea floor. Over time, countless eruptions cause the volcano to grow until it finally emerges above sea level to form an island. Continuing plate movement will eventually carry the island beyond the hotspot and volcanic activity will cease. The island will sink back into the ocean floor.

Vulcanicity

Volcanoes are built by the accumulation of their own eruptive products: **lava**, **volcanic bombs** (crusted over **ash** deposits), and **tephra** (airborne ash and dust). A volcano is most commonly a conical hill or mountain built around a **vent** that connects with reservoirs of molten rock below the surface of the earth. According to the European Space Agency, there are approximately 1500 active volcanoes around the world. Only a small proportion of them are erupting at any one time. An **eruption** is when a volcano gives off quantities of lava and volcanic gas. A few volcanoes (e.g. Mauna Loa, Hawaii) erupt more or less continuously, but others lie **dormant** between eruptions when they give out very little gas and lava. If a volcano has not erupted in the last 25 000 years it is labelled **extinct**.

Variations in the type and frequency of volcanic activity

Volcanoes occur in belts across the earth's surface following destructive and constructive plate boundaries (about 80% of the world's volcanoes are subduction volcanoes). Magma is produced where the mantle is rising along mid-ocean ridges and when crust is forced to sink and melt under oceanic trenches. There are also volcanoes that are found away from plate margins. These are associated with hotspots.

Fig 5
Global distribution of active volcanoes and earthquakes

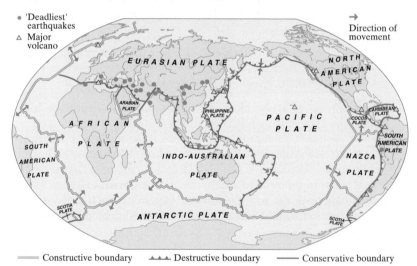

Examiners' notes

Using **fig 5** you should be able to describe the global distribution of active volcanoes and their relationship to plate boundaries.

The Pacific Ring of Fire is an almost continuous belt of volcanoes that comprises the Andes and Rocky Mountains (young fold mountains), and the island arcs stretching from the Aleutian Islands in Alaska to New Zealand.

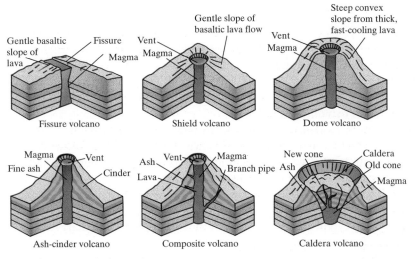

Fig 6
The main types of volcano

The type of volcano and volcanic activity depends upon the nature of the lava; this in turn depends upon the location of the volcano with regard to tectonic plate margins. If the lava is a thin fluid (not viscous), the gases may escape easily. But if the lava is thick and pasty (highly viscous), the gases will not move freely but will build up tremendous pressure, and ultimately escape with explosive eruptions.

The largest volcanoes are called **shield** volcanoes. These are found where the lava flows easily. They are built almost entirely of fluid lava flows. Flow after flow pours out in all directions from a central summit vent, or group of vents, building a broad, gently sloping cone with a flat, dome shape. Shield volcanoes include Kilauea and Mauna Loa on the island of Hawaii – two of the world's most active volcanoes. Mauna Loa, the largest of the shield volcanoes (and also the world's largest active volcano), projects 4 170 m above sea level and its volume is estimated at 80 000 cubic kilometres.

Some of the earth's grandest mountains are **composite** volcanoes (stratovolcanoes), e.g. Mount Fuji in Japan. They are typically large, steep-sided, symmetrical cones of alternating layers of lava flows, volcanic ash, cinders, blocks, and bombs, and may rise as much as 2 500 m above their bases.

Most composite volcanoes have a **crater** at the summit which contains a central vent or a clustered group of vents. Lava either flows through breaks in the crater wall or issues from fissures on the flanks of the cone. The essential feature of a composite volcano is a pipe system through which magma from a reservoir deep in the earth's crust rises to the surface.

Minor forms of extrusive activity

Geysers occur when hot water fills underground cavities. The water, upon further heating, is violently ejected when it suddenly flashes into steam. This cycle can be very regular – for example, Old Faithful Geyser in Yellowstone National Park, which erupts about once every 65 minutes.

Examiners' notes

Candidates must be able to link the types of volcano to their position with regard to plate boundaries – e.g. shield volcanoes are found away from plate boundaries. Include an explanation of type and location in your answers.

Essential notes

There is a difference between the type of volcano and the type of eruption. Task: Research a selection of eruption types, e.g. Hawaiian, Strombolian, Vulcanian, Pelean and Plinian. For each one make links to the location, type of lava and the resultant type of volcano.

This topic continues on the next two pages

Hot springs occur where groundwater is constantly heated, but not boiled, by hot rocks beneath the surface.

Boiling mud pools form where steam and gas rise to the surface under rainwater ponds. The acidic gases attack surface rocks, forming clay. The clay-rich soil mixes with the pond water to produce a muddy, steam-heated slurry, or mud pool. They are common in North Island, New Zealand.

Forms of intrusive activity

Sometimes, magma never reaches the surface. It is intruded into weaknesses in rocks and then cools down. When the rocks above them are eventually eroded away, the intrusions often make prominent landmarks. **Dykes** are steep, sheet-like intrusions, varying in thickness from a few millimetres to tens of metres across. They occupy vertical weaknesses in the rock. They often cut across rock bedding and form low ridges. **Sills** are sheets of **igneous** rock that follow the bedding of sedimentary rock layers. The Great Whin Sill in northern England is a good example. Deep-seated masses of granitic magma may rise as huge blobs, like oil in water, to solidify within a few kilometres of the surface to form plutons. These plutons may be added to over millions of years to form **batholiths**, some reaching over 1 000 km in length. A good example of this is the batholith underlying southwest England running from Dartmoor to Land's End.

Volcanoes as a hazard

A natural **hazard** is the potential threat to humans from a naturally occuring process or event.

A volcano is only deemed hazardous when its activity has an impact on people. Some of this activity will only have a localized effect (e.g. bombs), others will have a more regional impact (e.g. **lahars**). The eruption cloud could have global effects as fine ash reaches the high atmosphere and circulates around the globe.

Case study of a recent volcanic event: Eyjafjallajökull

Eyjafjallajökull is one of Iceland's smaller ice caps located in the far south of the island. The ice cap covers the caldera of a volcano 1 666 m in height.

The volcanic events starting in March 2010 are considered to be a single eruption divided into different phases. Initially a fissure opened up, about 150 m in length running in a northeast to southwest direction, with 10 to 12 erupting lava craters ejecting lava at a temperature of about 1 000 °C up to 150 m into the air. The lava was basalt and relatively viscous, causing the lava stream to be slow. The molten lava eventually flowed more than 4 000 m to the northeast of the fissure. Ash ejection from this phase of the eruption was small, rising to no more than 4 km into the atmosphere. On 14 April 2010, the eruption entered an explosive phase and ejected fine, glass-rich ash to over 8 km into the atmosphere, which was then deflected to the east by westerly winds. The **second phase** is estimated to have been a **volcanic explosive index** (VEI) 4 eruption, which is large, but not nearly the most powerful eruption in VEI terms.

Essential notes

Take an example of each of the following: dyke, sill and batholith. For each example, make notes on how it affects the landscape.

Examiners' notes

The specification states that candidates should undertake **two** case studies of recent (ideally within the last 30 years) volcanic events. They should be taken from contrasting areas of the world. In each case, the following should be examined:

- The nature of the volcanic hazard
- The impact of the event
- Management of, and responses to, the hazard

Impacts

The volcano is in a remote corner of Iceland. The local, dispersed farming community is home to few people. There were minor local impacts:

- A thick layer of ash fell on farm pastures at Raufarfell. This has become wet and compact, making it very difficult to continue farming, harvesting or grazing livestock.
- Locally, river levels rose as part of the ice cap melted.
- Some local gravel roads were blocked by falling ash.

What made this volcanic activity so hazardous was the fact that the ash became disruptive to air travel. This was due to a combination of the following four factors:

1. The volcano's location is directly under the polar jet stream.
2. The direction of the jet stream was unusually stable at the time of the eruption's second phase. It maintained a continuous NW to SE heading.
3. The second eruptive phase took place under 200 m of glacial ice. The resulting meltwater flowed back into the erupting volcano which created two specific phenomena:
   - The rapidly vaporizing water significantly increased the eruption's explosive power.
   - The erupting lava cooled very rapidly, which created a cloud of highly abrasive, glass-rich ash which could damage jet engines.
4. The volcano's explosive power was sufficient to inject ash directly into the jet stream over NW Europe.

Responses

- Five hundred local farmers were evacuated overnight.
- Some roads were closed because of a fear of flash floods.
- Between 14 and 21 April flights in the airspace of many NW European countries were stopped. There was then sporadic disruption to flights for several days depending on the varying intensity of ash cloud and weather patterns. The six-day shutdown is estimated to have cost the airlines £1.2bn.
- A Royal Navy warship collected soldiers returning from Afghanistan and stranded holidaymakers from the Spanish port of Santander.

Examiners' notes

Examiners are looking for precise case study detail to support arguments. Factual detail is important because it can support discussion. Even three or four hard facts about the impacts of a particular hazard will sound convincing.

Ash cloud

Fig 7
The extent of the ash cloud on 15 April 2010

Examiners' notes

Produce a case study of a volcano from a contrasting part of the world. Follow the same plan as in this text so that comparisons can be made.

Seismicity

The causes and main characteristics of earthquakes

As the crust of the earth is mobile, there tends to be a slow build-up of stress within the rocks. When this stress is suddenly released, parts of the surface experience an intense shaking motion that lasts for between a few seconds and a few minutes. This is an earthquake.

The point of pressure release is the **focus** and the point immediately above that on the earth's surface is the **epicentre**. The depth of focus is significant:

- Shallow earthquakes (0–70 km) cause the most damage, with 75% of all energy released.
- Intermediate (70–300 km) and deep (300–700 km) earthquakes have much less effect on the surface.

Earthquakes originate along **faults.** Parts of the crust are being forced to move in opposite directions. These huge masses of rock get stuck, but the forces on them continue, building up stresses in the rocks. Eventually the strain overcomes friction and the rocks move, releasing large amounts of energy. This energy is transferred to the surrounding rocks, travelling through them as **seismic waves**. A lot of the energy is transferred vertically to the surface and then moves outwards from the epicentre.

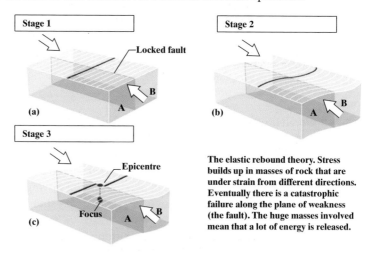

The elastic rebound theory. Stress builds up in masses of rock that are under strain from different directions. Eventually there is a catastrophic failure along the plane of weakness (the fault). The huge masses involved mean that a lot of energy is released.

Fig 8
The elastic rebound theory of earthquake generation

Seismic waves: There are several different kinds of seismic waves, and they all move in different ways. The two main types of waves are body waves and **surface waves**. Body waves can travel through the earth's inner layers, but surface waves can only move along the surface of the planet like ripples on water. Earthquakes radiate seismic energy as both body and surface waves. Most of the shaking felt from an earthquake is due to the **Rayleigh wave**, which can be much larger than the other waves. These were named after John William Strutt, Lord Rayleigh, who mathematically predicted the existence of this kind of wave in 1885. A Rayleigh wave rolls along the ground just like a wave rolls across a lake or an ocean. Because it rolls, it moves the ground up and down as well as side to side, in the same direction that the wave is moving.

Distribution of earthquakes

- Most earthquakes occur along plate boundaries or deep under continents (see **fig 5**). The tectonic conditions in which earthquakes occur are shown in **fig 9**.

Examiners' notes

Candidates must be able to describe global distributions. In this case, the distribution must relate to the type of plate boundary, the depth of the earthquake and the frequency of the earthquakes.

(a) Earthquakes occur along constructive plate boundaries where the upwelling magma forces the oceanic plates apart and faulting occurs on the edge of the central rift valley

(b) Earthquakes also occur along the transform faults that link offset parts of the rift valley

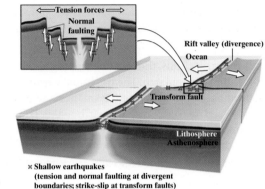

× Shallow earthquakes
(tension and normal faulting at divergent boundaries; strike-slip at transform faults)

(c) Increasingly deeper earthquakes occur underneath young fold mountains as the plate descends. This is called the Benioff zone

Fig 9
Faulting and earthquakes along constructive and destructive plate boundaries

Measurement of earthquakes

The **moment magnitude scale** (abbreviated as **MMS**, denoted as M_w) is used to measure the size of earthquakes in terms of the energy released. Earthquakes of $M_w 2$ or less are rarely felt by humans. The scale is logarithmic. An increase of 1 unit of magnitude increases the amount of shaking by 10, but the amount of energy released by 30.

Essential notes

The modified Richter magnitude scale is often used in the media to describe the intensity of earthquakes even though it is the moment magnitude number that is quoted.

Faulting and earthquakes along destructive plate boundaries

Case study 1: the Sichuan earthquake, 2008

On Monday 12 May, 2008 at 14:28 (local time), an earthquake of magnitude 7.8 struck the province of **Sichuan** in southeastern China.

This part of China is affected by the collision between two tectonic plates. The Indo-Australian Plate is moving northwards at about 5 cm a year, into the Eurasian Plate. There is convergence between the Tibetan plateau and the Sichuan Basin. On 12 May 2008 stresses in the crust triggered a sudden movement along a thrust fault on the northwest margin of the Sichuan Basin releasing powerful earth tremors across the region.

The earthquake occurred in two stages: the 250-km long Longmenshan Fault tore in two sections, the first one ripping about 7 m, followed by a second one that sheared 4 m. The earthquake lasted about two minutes. The shallowness of the epicentre (19 km) and the density of population greatly increased the severity of the earthquake. The seismic waves of the quake travelled a long distance without losing their power because of the firmness of the terrain in central China. In the 72 hours following the main earthquake there were between 64 and 104 major **aftershocks** recorded, ranging in magnitude from 4.0 to 6.1. Some of these were responsible for further deaths.

Impacts:

- Official figures state that 69 197 were confirmed dead, including 68 636 in Sichuan province alone. The earthquake left about 4.8 million people homeless, though the number could be as high as 11 million.
- In terms of school casualties, thousands of school children died due to shoddy construction. In Mianyang city, seven schools collapsed, burying at least 1 700 people. At least 7 000 school buildings in Sichuan province collapsed. Up to 1 300 children and teachers died at Beichuan Middle School.
- There was significant damage to agriculture; in the Sichuan province alone, a million pigs died.

Fig 10
Building damage caused by the 2008 Sichuan earthquake

- Landslides and rockfalls destroyed communications, making it difficult for rescuers to get into this mountainous region. Landslides also killed 700 people in Qingchuan and a train was buried by a landslide near Logan, Gansu province.
- Thirty-four barrier lakes formed due to earthquake debris damming rivers and it was estimated that 28 of them were a potential danger to the local people.

Management of the hazard:

- Just 90 minutes after the earthquake, Premier Wen Jiabao, who has an academic background in geomechanics, flew to the earthquake area to oversee the rescue work.
- Eighty thousand troops were mobilized almost immediately. One of their first tasks was to deal with rising water levels on the largest barrier lake (Tangjiashan lake on the Jian river) which threatened to breach a temporary earth dam. A quarter of a million people were evacuated downstream to higher ground. Eventually an artificial channel, completed on 7 June, drained the lake.
- The Chinese government believed that many areas were now too dangerous for people to live in. A programme of relocation of mountain communities is being carried out.
- Priority has been given to building nearly 4 million new homes, creating 1 million new jobs, and constructing high-quality buildings that are earthquake-proof. Three million homeless rural families will get new houses, and 860 000 city apartments will be built.
- The Chinese government immediately committed £800m to strengthening the 2 600 schools that remained standing. The reconstruction plan also included 169 new hospitals and nearly 4 500 new primary schools to be built in Sichuan and neighbouring Gansu and Shaanxi provinces.
- Welfare programmes have been expanded to help the 1.4 million people driven to poverty by the disaster. The ultimate aim is to create an earthquake-resistant society, making Sichuan less vulnerable.

Responses to the earthquake:

- The Chinese government were praised by the international community for their prompt and efficient relief effort.
- China opened up to international aid organizations.

However:

- Bereaved parents were outraged because of the many schools that had collapsed killing their only child. Studies have found that the schools had been built quickly and many earthquake protection measures had not been built in to the design. For people who have lost a child, the one-child policy has been suspended to allow parents to have another.
- There have been protests about the school architecture, but these have been suppressed by the Chinese government.

The total bill for reconstruction will be $150bn.

Examiners' notes

The issue of the loss of children, and how it relates to China's population policies, is an example of where different parts of the specification overlap. This is a good example of synopticity.

This topic continues on the next two pages

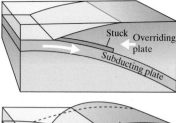

Earthquake starts tsunami

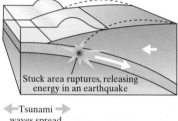

Stuck area ruptures, releasing
energy in an earthquake

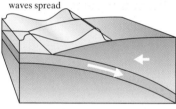

Fig 11
How the Indian Ocean tsunami was
generated

Case study 2: the Sumatran-Andaman earthquake, 2004

This was primarily known for the generation of the great Indian Ocean **tsunami.**

The nature of the seismic hazard: The Indo-Australian Plate, which underlies the Indian Ocean and Bay of Bengal, is drifting northeast at an average of 5 cm/year. Where this meets the Burma Plate it subducts beneath the Burma Plate. This has formed the Sunda trench and the Sunda island arc.

The earthquake was unusually large in both geographical and geological extent. An estimated 1 600 km of fault surface slipped about 15 m along the subduction zone. The slip did not happen instantaneously but took place in two phases over a period of several minutes. Data indicate that the first phase involved a rupture about 400 km long and 100 km wide, located 30 km beneath the seabed. This was the largest rupture ever known to have been caused by an earthquake. The sea floor is estimated to have risen by several metres, displacing an estimated 30 km³ of water. The waves did not originate from a point source, but rather radiated outwards along the entire 1 600-km length of the rupture. This greatly increased the geographical area over which the waves were observed.

Out in the open ocean, the tsunami wave has a very long wavelength and travels at high speeds; when the ocean is 6100 m deep, a tsunami will travel about 890 km/hr. As a tsunami leaves the deep water and arrives at shallow coastal waters, the velocity of the tsunami decreases. The frequency of the wave remains the same so the height of the wave increases.

Scientists investigating the damage in Aceh, the nearest landfall to the epicentre, found evidence that the wave reached a height of 24 m when coming ashore along large stretches of the coastline, rising to 30 m in some areas when travelling inland.

The northern regions of the Indonesian island of Sumatra were hit very quickly, while Sri Lanka and the east coast of India were hit roughly 90 minutes to two hours later. Thailand was also struck about two hours later despite being closer to the epicentre, because the tsunami travelled more slowly in the shallow Andaman Sea off its western coast.

The tsunami was noticed as far away as Struisbaai in South Africa, some 8 500 km away, where a 1.5 m high wave surged on shore about 16 hours after the earthquake.

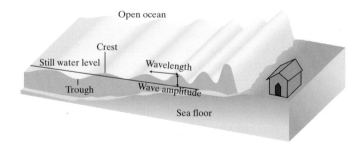

Fig 12
How a tsunami steepens as it
approaches a coastline

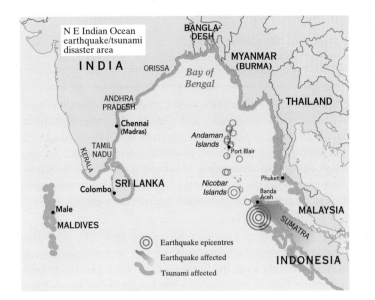

N E Indian Ocean earthquake/tsunami disaster area

INDIA · ORISSA · ANDHRA PRADESH · Chennai (Madras) · TAMIL NADU · KERALA · Colombo · SRI LANKA · Male · MALDIVES · BANGLA-DESH · MYANMAR (BURMA) · Bay of Bengal · THAILAND · Andaman Islands · Port Blair · Nicobar Islands · Phuket · Banda Aceh · MALAYSIA · SUMATRA · INDONESIA

◎ Earthquake epicentres
Earthquake affected
Tsunami affected

Fig 13
Countries affected by the Indian Ocean tsunami

The impacts of the Indian Ocean tsunami

- Of the 236 210 estimated deaths, 184 167 have been confirmed.
- With an estimated death toll of 167 736, Indonesia was the worst affected. Sri Lanka was able to confirm that 35 322 people died. India and Thailand were also badly affected with 18 045 and 8 210 deaths respectively.
- There were 125 000 people injured and a further 45 752 people are missing.
- 1.69 million people were displaced.
- Up to 9 000 foreign tourists were killed, many from northern Europe. Sweden alone had 540 fatalities.
- The economy of the coastal communities was badly hit. Many fishermen were killed. Of those who survived many lost their boats.
- The tourist industry was badly hit; this was the peak of the travel season to this part of the world. Not only was a lot of the tourist infrastructure damaged, but many people have been put off travelling to that part of the world. This has had a drastic effect on one of the main sources of foreign earnings for the area.
- The salt water incursion onto the land poisoned soil. Sewage systems and drains were destroyed and water-borne diseases became a major problem. There has been an increase in the incidence of malaria because the sterile pools of water allowed mosquitoes to breed.
- It is estimated that the total rebuilding costs across the area will amount to well over £8bn.

Essential notes

The management of this disaster was very different to that of the Sichuan earthquake. There were more countries affected and for the most part they were neither as wealthy nor as organized as the Chinese government. Overseas aid was an essential part of the management effort.

Task: Take one place affected by the tsunami. For that place, outline how the relief effort was carried out.

Major climatic controls

The structure of the atmosphere

The **atmosphere** is a mixture of gases. These are often divided up into the major constant components, and the highly variable components. Dry air contains approximately (by volume) 78.09% nitrogen, 20.95% oxygen and 0.93% argon. These are constant over time and location and have little effect on the variations in **weather** and **climate**. The variable components (e.g. carbon dioxide, sulphur dioxide, methane and ozone) make up less than 1% of atmospheric gases. They affect long-term climate change because they are **greenhouse gases**, keeping the planet warm. Water vapour varies between 1% and 4%. This has a huge effect on differences in humidity. The atmosphere also contains particulate matter such as dust, volcanic ash, salt and smoke.

The atmosphere is made of four main layers (see **fig 1**). The most important layer, as far as weather and climate is concerned, is the **troposphere.** This is the lowest layer and varies in thickness between 8 km and 12 km. It contains 75% of the atmospheric gases. The temperatures drop by approximately 6.5 °C with every 1 000 m of altitude. All the earth's weather takes place in the troposphere. At the top of the troposphere is a boundary called the **tropopause**. This is where strong high-level winds, called **jet streams**, flow.

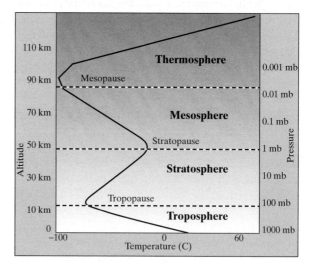

Fig 1
The vertical structure of the atmosphere

The atmospheric heat budget

The atmospheric **heat budget** is the balance between the incoming solar radiation (**insolation**) and outgoing radiation from the planet. The sun is the source of energy that drives the atmospheric engine. **Fig 2** shows what happens to the solar energy. Forty-nine per cent of this is absorbed by the clouds and atmosphere or reflected by them and the earth's surface. The 51% of the solar energy that reaches the surface is radiated back to space (6%) or used to heat the atmosphere in some way (conduction, **latent heat** and absorption) (45%). The amounts given in **fig 2** vary considerably. Two examples of this variability are as follows:

- In desert regions, where there is a clear sky, there is no cloud reflection and so more energy reaches the surface. On the other hand, the clear skies at night, and the low humidity, allow large amounts of direct radiation into space.
- In the ice-covered polar regions, almost all the energy is reflected from the surface. There is very little heating of the air.

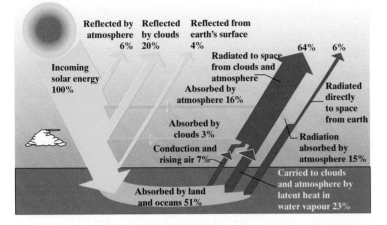

Fig 2
The general atmospheric heat budget

Global temperatures have remained constant (with minor fluctuations) for the last 5 000 years. This means that the amount of incoming radiation must be the same as the amount of outgoing radiation. In recent years, with the increase of greenhouse gases, the atmosphere is able to absorb more of the outgoing radiation.

This is complicated by the fact that the net amount of heat absorbed and radiated varies with **latitude**. **Fig 3** shows that the area between 40 °N and 40 °S has a net gain in heat, while the higher latitudes have a net loss. However, the equatorial regions are not getting any hotter, and the poles are not getting any colder. To balance the budget, heat is transferred from the equatorial areas to the poles by wind (air movements that include **trade winds**, jet streams, **depressions** and tropical storms) and by water movement (ocean currents). Heat is also transferred vertically from the lower atmosphere by conduction, convection and latent heat (which is used in the change of state from water to water vapour).

Essential notes

The atmosphere gains much more heat energy from the earth's surface than it gets from the sun. *'The sun heats the earth, the earth heats the atmosphere.'*

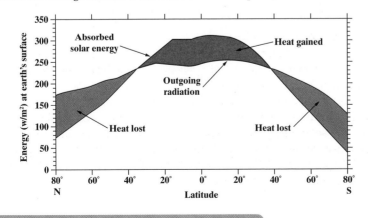

Fig 3
Heat budget changes with latitude

This topic continues on the next two pages

General atmospheric circulation

Atmospheric pressure is the force per unit area exerted against a surface by the weight of air above that surface. Low pressure areas have less atmospheric mass above their location, whereas high pressure areas have more atmospheric mass above their location. Atmospheric pressure reduces with altitude because there is less air above the measuring point. Air pressure also varies with air temperatures. Warm air has fewer molecules per cubic metre than cold air and so exerts less pressure on the surface below. Understanding atmospheric pressure helps us understand global air circulation as summed up in the **tri-cellular model** below.

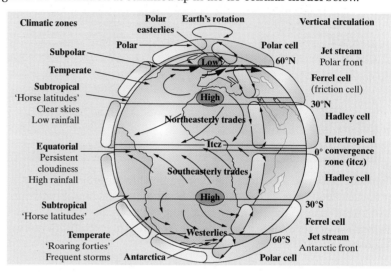

Fig 4
The tri-cellular model of atmospheric circulation

The tri-cellular model

The Hadley cell: In the equatorial regions, with the intense heating, air rises by convection and produces a low pressure region. As this air reaches the top of the troposphere, it begins to move polewards. It eventually cools and sinks (exerting high pressure) at around 30 °N and S. These are called the subtropical high pressure regions. When the falling air reaches the surface some of it returns to the equator and some moves polewards.

The polar cell: the result of high pressure cold air, descending over the poles and then blowing outwards. These winds eventually warm up and at the low pressure belt at approximately 50–60 °N and S (the temperate low pressure zone) they rise and either return to the pole or flow towards the subtropics.

An intermediate cell, the **Ferrel cell**, lies between the Hadley and polar cells.

Planetary surface winds

Air blows from regions of high atmospheric pressure to regions of low atmospheric pressure. This causes the winds that blow from the subtropics to the equator and the temperate low pressure zones. These winds do not blow directly north or south because they are apparently deflected to the right as the earth spins beneath them (the **Coriolis force**). This gives rise to the trade winds and the westerlies.

The effect of latitude

The latitude affects climate in two ways:

1. The angle of incidence of the sun is lower as you move away from the equator. This means that towards the poles:
    *   A larger surface area is covered by the same amount of insolation, thus less energy per unit area
    *   There is greater reflection of sunlight
    *   There is more atmosphere for the sunlight to pass through, resulting in greater atmospheric scattering and reflection

2. The length of daylight hours. This affects the amount of insolation an area receives: the longer the daylight, the greater the insolation. In the tropics this is constant at 12 hours per day. In summer, polar regions have up to 24 hours of daylight, and no daylight at all in the middle of winter.

The effect of oceanic circulation

Ocean currents are set in motion by the wind. The rotation of the earth moves them clockwise in the northern hemisphere and anticlockwise in the southern hemisphere.

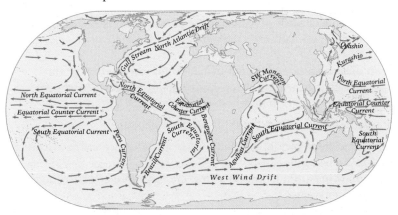

Fig 5
Oceanic circulation

Warm currents flow along the western sides of oceans and cold currents flow along the eastern sides. Warm currents work against the effects of latitude. Air that blows over a warm current warms up. When this blows onshore it raises the temperatures. Cold currents do the opposite, cooling air down. This has the further effect of creating coastal deserts in subtropical regions (e.g. the Atacama Desert, South America).

The effect of altitude

Temperatures decrease with height above sea level. Globally the reduction in temperature with height (the environmental **lapse rate**) is approximately 6.5 °C per 1 000 m. If dry air is forced to rise (e.g. over a mountain), the consequent expansion in volume causes a temperature fall of 10 °C per 1 000 m. This is called the dry adiabatic lapse rate (DALR). If the air is saturated and condensation is taking place there is a release of latent heat. This reduces the rate of cooling to 5 °C per 1 000 m. This is called the saturated adiabatic lapse rate (SALR).

Essential notes

The changes in temperature of gases (in this case the atmosphere) caused by changes in pressure and volume alone are called **adiabatic** changes.

The climate of the British Isles

Basic climatic characteristics

Temperature: The British Isles lie approximately between latitudes 50 °N and 60 °N. This locates them in the **temperate** low pressure region at the boundary between the polar and Ferrel cells. The temperate nature of the climate is further moderated by the maritime position of the British Isles and the warm **North Atlantic Drift** ocean current. This all gives a pattern where there are no extremes of temperature.

Precipitation: The average annual rainfall varies enormously over the British Isles from about 5 000 mm in parts of the Western Highlands of Scotland, to about 500 mm in parts of East Anglia and the Thames estuary. Overall, the wettest areas are in the western half of the country. The reasons for this are as follows:

- The most common (**prevailing**) winds are from the southwest and south. They are moist from blowing over the Atlantic Ocean.
- The western side of the British Isles has the highest relief. The moist air rises over the high relief and produces relief (**orographic**) rainfall on the mountains and a rain shadow to the east.
- The major cause of rainfall in the British Isles is **depressions**. These mainly approach from the west.

Air masses

An **air mass** is an extensive body of air in which there is only gradual horizontal change in temperature and humidity at a given height. Air masses result from the nature of the surface and the latitude.

Air masses resulting from the nature of the surface are:

- **Continental (c)** (of the land) – these are often dry and have extremes of temperature.
- **Maritime (m)** (of the sea) – these are very moist and have moderate temperatures.

Air masses resulting from latitude are:

- Arctic (A): this is very cold air.
- Polar (P): this is cold air.
- Tropical (T): this is warm/hot air.

The air masses that affect the British Isles are:

- Tropical maritime (Tm): this is the most common air mass over Britain. It brings mild wet weather in winter and warm damp weather in summer.
- Polar maritime (Pm): this is very common over Britain. It brings cool conditions with rain in summer and snow to NW Scotland in winter.
- Polar continental (Pc): this is characterized by cold winter temperatures. It can bring snow to the east of Britain in winter. In summer this air brings warm conditions that can last for some time.
- Tropical continental (Tc): this only affects the British Isles in summer. It can bring heatwaves and cause thunderstorms.

- Arctic maritime (Am): this brings very cold weather in winter (−10 °C), snow in northern Scotland. It rarely affects the British Isles in summer.

The origin and nature of depressions

A depression is an area of low atmospheric pressure. The British Isles lie in the latitude of prevailing westerly winds where depressions and their associated fronts (bands of cloud and rain) move eastwards or north-eastwards across the North Atlantic, bringing with them unsettled and windy weather, particularly in winter.

A front is a boundary between different air masses.

The polar front occurs where the cold air from the **polar cell** meets the warm air from the **Hadley cell** (see the tri-cellular model, **fig 4**). This front is not a straight line and waves appear in it. Where the waves extend north there is low atmospheric pressure. This can lead to 'kink' in the polar front which travels along the front.

The British Isles sit right in the path of the polar front and so our weather is dominated by these depressions. This means that the British Isles experiences polar air, then tropical air and then back to polar air again. When warm air rises over cold air it is called a **warm front**. When the colder polar air pushes under the warmer tropical air then this is called a **cold front**.

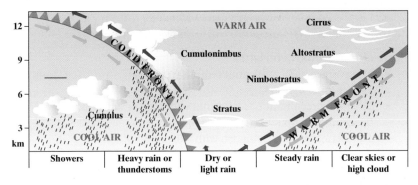

Rain occurs along fronts because the air is rising. Rising air cools because it expands. Cooler air can hold less water vapour. Water vapour condenses to form clouds. Cloud droplets coalesce to form raindrops.

The origin and nature of anticyclones

Anticyclones are areas of relatively high pressure. They move slowly and may remain stationary over an area for several days or even weeks.

The air in an anticyclone falls and warms due to increased atmospheric pressure. This causes evaporation of water droplets which in turn leads to cloudless skies. Winds are weak and flow outwards in a clockwise direction.

In winter, anticyclones bring cold days, very cold nights, ground frost and hoar frost, and radiation fogs. In summer, anticyclones bring hot days, warm nights, clear skies, early morning mist and dew and thunderstorms.

The climate of tropical regions: tropical wet/dry savanna

This climate is found throughout the tropics. An example of an area that experiences a **savanna** climate is west Africa. This climate is considered to be transitional between the equatorial and desert climates. Bamako in Mali is a good example of this type of climate.

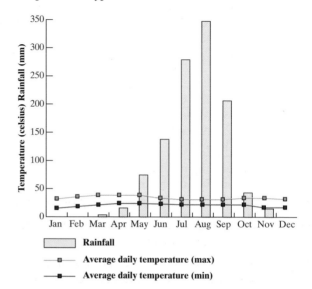

Fig 7
The average yearly climate of Bamako, Mali

Essential notes

North of Bamako:

- It is hotter in summer, cooler in winter.
- It is drier, with a longer dry season.
- The wet season is shorter and the rainfall is less reliable.

Task: Research what happens to the savanna climate to the south of Bamako.

Basic climatic characteristics

- The average daily maximum **temperatures** range from 31°C (Aug) to 39°C (March to May).
- The total precipitation is 1 100 mm with four months drought (December to March).
- The wettest month is August, with 348 mm rainfall.
- Because the temperature remains high through the year, it is known as a wet/dry season climate.

The winds in west Africa are also seasonal. In the wet season, the winds are from the SE. In the dry season they are from the NE.

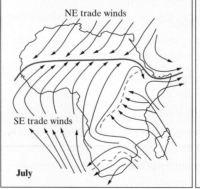

Fig 8 Africa's trade winds, July

Fig 9 Africa's trade winds, January

──── Intertropical convergence zone
───► Wind currents

The savanna climate of west Africa is determined by its latitude. In the tri-cellular model of atmospheric circulation (**fig 4**) it can be seen that west Africa lies between two pressure zones:

- The **subtropical high pressure zone** where the falling air is very dry. This has a desert climate.
- The low pressure equatorial region where there is rising air. This is called the **intertropical convergence zone (ITCZ)** because it is located where the **NE trade winds** meet (converge with) the **SE trade winds**.

The low pressure ITCZ migrates throughout the year. **Figs 8** and **9** show its location in west Africa in July and January. This migration is, in part, due to the migration of the overhead sun. The higher the sun is in the sky, the greater the heating effect.

In June, the sun is overhead at the Tropic of Cancer (23.5°N). The ITCZ migrates northwards. It can reach as far north as 20°N but often does not. By December, the overhead sun has moved to the Tropic of Capricorn (23.5°S). The ITCZ follows the overhead sun in east Africa, but because of the shape of the continent and the fact that the land heats up more than the sea, the ITCZ remains at between 5°N and 10°N in west Africa.

The ITCZ draws in both hot dry air from the subtropical desert and hot moist air from the warm Atlantic Ocean.

In January, this moist air is drawn onto the coastal area of west Africa. As it reaches the ITCZ it rises, and heavy **convectional rainfall** occurs. To the north of the area, air is drawn southeastwards from the subtropical high pressure area of the Sahara desert. This is dry and dusty. It is a regular occurrence and is called the Harmattan. This evaporates any moisture and is responsible for the dry season and drought of the inland part of west Africa.

As the ITCZ migrates northwards, moist air is drawn further north, causing the belt of rain to move with the ITCZ. By July, the ITCZ reaches its northern limit. The area north of approximately 15°N is not always reached and so sometimes the northerly parts of the region have years with very little rainfall. This region, known as the **Sahel**, suffers regular droughts when the rains fail.

The unpredictable nature of the rainfall in the northern savanna impacts upon the people living there. Droughts lead to crop failure and starvation.

Recent work by meteorologists and oceanographers has shown that much of the recent year-to-year changes in Sahel rainfall are forced by changes in sea-surface temperature in the Gulf of Guinea. When the Gulf of Guinea is warm, the ITCZ shifts south away from the Sahel, reducing the African monsoon that draws moist air into the Sahel. This oceanic forcing of Sahel rainfall is exacerbated by land–atmosphere feedbacks. As the land dries out, there is less vegetation, less evaporation from the land, and more sunlight is reflected from the land. These processes further weaken the monsoon.

Examiners' notes

You should show an understanding of the processes that lead to convectional rainfall.

Tropical revolving storms

A **tropical revolving storm** is the generic term for a low pressure system over tropical or subtropical seas and oceans, with organized convection (i.e. thunderstorm activity) and winds at low levels, circulating either anticlockwise (in the northern hemisphere) or clockwise (in the southern hemisphere).

They are defined as an intense rotating depression which develops over tropical oceans, and has winds exceeding 119 k/h. On average there are 80 of these storms each year with two thirds of them occurring in the northern hemisphere.

Tropical revolving storms require a number of conditions to develop:

- A source of warm, moist air derived from tropical oceans with sea surface temperatures normally in the region of, or in excess, of 27 °C
- Winds near the ocean surface blowing from different directions converging and causing air to rise and storm clouds to form
- Winds which do not vary greatly with height, known as low wind shear; this allows the storm clouds to rise vertically to high levels
- Spin induced by the rotation of the earth (the Coriolis force); they are not able to form within 500 km of the equator because this force is not large enough.

Fig 10
Global distribution of tropical revolving storms

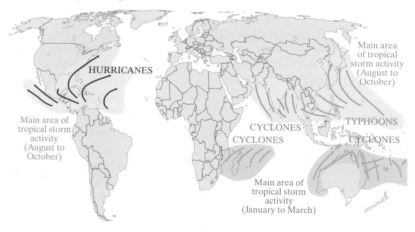

There is a high evaporation rate from the warm oceanic water. This moist air rises by convection. As it rises the temperature falls until the air is saturated. Condensation occurs and latent heat is given off. This latent heat fuels further uplift. If these conditions develop into a full tropical revolving storm they are given a category depending on the winds blowing around them. The Saffir-Simpson intensity scale is used.

The average tropical revolving storm is 320 to 500 km in diameter, though massive tropical revolving storms can span 1 100 km or more.

Tropical revolving storms produce strong sustained winds with even stronger gusts (over 280 k/h), very heavy rainfall (over 300 mm in a very short time), localized tornadoes (small short-lived revolving storms) and large storm surges.

The impact of a tropical revolving storm depends upon the intensity of its low pressure, its scale and its location. Tropical cyclones can remove forest canopy, shape coastal areas, and lead to mudslides and landslides in mountainous areas. After the storm, standing water can cause the spread of disease. Transportation or communications infrastructure may be destroyed, hampering clean-up and rescue efforts. Nearly 2 million people have died globally due to tropical revolving storms.

Case study of a tropical revolving storm: Cyclone Nargis, 2008

Occurrence: In early May 2008, Cyclone Nargis passed over Myanmar (Burma) after forming in the Bay of Bengal. Even though the storm lost strength before coming ashore on 2 May it still carried very powerful winds and heavy rain. The storm's path took it close to Yangôn, a city with a population of more than 4 million.

Physical impacts: The storm hit part of the Irrawaddy delta. This is a vast expanse of flat, low-lying land used to grow paddy rice. A storm surge of 4 to 5 m high flooded land up to 40 km inland by flowing up the Irrawaddy distributaries. Much of the remaining mangrove forest was destroyed, making the country more vulnerable in the future.

The effect of the storm surge was made much worse by the fact that a natural buffer zone of inter-tidal mangrove forest has been extensively cleared to make way for fish and shrimp farming.

Social and economic impacts: The United Nations estimates that in total 2.4 million people in Myanmar were affected by Cyclone Nargis. A total of 77 738 people died and 55 917 were missing. Approximately 150 000 people were displaced into about 120 temporary settlements.

Cyclone Nargis also wrecked as much as 65% of Myanmar's rice crop – at least 200,000 ha of the Irrawaddy delta were ruined. The storm hit just a few days after the harvest was completed and so also wiped out much of the crop in warehouses.

In the capital Yangôn (Rangoon), the cyclone knocked out power and water supplies, felled trees and damaged hundreds of buildings. Even so, many people who live on the outskirts of Yangôn lost their houses and had to move to temporary shelters.

Response: For the first month, Myanmar's military leaders, fearful of external influence, refused to let international aid groups into the country.

Since then, more than $450m was committed to the region, but it is less than half of the $1.1 billion called for by the UN.

Examiners' notes

Storm surges are one of the most destructive aspects of tropical revolving storms. Candidates must be able to explain what they are.

Essential notes

Care International is a charity that worked in Myanmar following Cyclone Nargis.

Find out what they did at www.careinternational.org.uk

Examiners' notes

The specification requires candidates to have two case studies of tropical revolving storms from contrasting areas of the world. Follow the same pattern as has been done for Cyclone Nargis so that comparisons can be made.

Climates on a local scale: urban climates

Urban climates are any set of climatic conditions that prevail in large metropolitan areas and that differ from the climate of their rural surroundings.

Temperature

The **urban heat island** effect (UHIE) is the most studied of the urban climatic effects. It refers to the generally warm urban temperatures compared to those of surrounding, non-urban, areas. Initially, the effect was studied because of its warming impact on cities in winter. Increasingly, scientists are studying the effect in summers when potentially lethal heat-waves can occur.

The UHIE is the product of a variety of factors. These include the following:

- **Anthropogenic** sources of heat. These include heat given off by people, machines, heating escaping from buildings, air-conditioning systems, industrial processes and cars.
- Multiple reflections of incoming solar radiation from tall buildings that enable absorption to take place on more than one surface.
- Urban surfaces tend to have lower **albedos** (a measure of reflectivity), which enables them to absorb more of the incoming solar radiation. This is combined with the higher heat capacity of urban surfaces, which allows them to absorb the heat and store it. This is released when the air begins to cool at night.
- The efficient drainage of the urban surface removes surface water quickly. There is less capacity for evaporation to take place with its associated cooling effect.
- There are lower amounts of vegetation which cool the air by transpiration.
- Above many cities there is a dome of **particulate** and NO_2 pollution. This allows the shortwave radiation from the sun into the urban atmosphere. It then absorbs and reflects the outgoing longer wave radiation, preventing its escape.
- The increased cloud amount over the urban area also reflects outgoing radiation back to the surface.
- The rough urban surfaces reduce the wind speed and its ability to flush out the warm air.

The UHIE could cause problems in the future. Over 3 billion people now live in cities; by 2030 it could be up to 5 billion people. With little access to air conditioning, refrigeration, or medical care, the world's urban poor will be particularly vulnerable to heatwave-related health hazards.

Precipitation

Urban areas tend to have up to 10% more rainfall than surrounding areas. Rain lasts for longer and intense storms with thunder are more common.

The increase in rainfall is because:

- The UHIE causes convectional uplift and consequent convectional rainfall. This also increases the number of thunderstorms and the

Essential notes

Research the albedo values of common urban surfaces, compared to common rural surfaces.

Examiners' notes

Case studies are important in extended essays because they illustrate what is actually happening. Research at least one case study of a well-known urban heat island (e.g. Atlanta Ga., USA).

average intensity of the rainfall.

- The presence of high-rise buildings can cause turbulence which can lead to uplift of air.
- Particulate pollution means that there is an increased number of hygroscopic nuclei present in urban air.

Fogs: In urban areas, fogs are 100% more frequent in winter and 30% more frequent in summer than in the surrounding rural areas. Urban fogs are the result of a greater concentration of air-borne particulates that act as condensation nuclei. Fogs are most common near urban rivers. These are often artificially warmed by effluent from industry and so have high evaporation rates.

Air quality

Particulate pollution: The notation PM_{10} is used to describe particles of 10 micrometres or less. These are the result of human activity, particularly industrial processes and vehicle exhausts. They can include cement dust, tobacco smoke, ash and coal dust.

Photochemical smog: This is the result of a chemical reaction between sunlight and nitrogen oxides (NO_x). This can be the cause of health problems (headaches, eye irritation, coughs and chest pains) as well as being damaging to vegetation.

Wind speed

Urban areas reduce average wind speed by increasing the friction between urban surfaces and moving air.

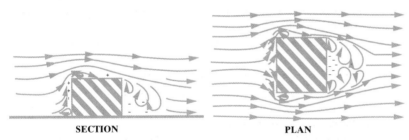

SECTION   PLAN

There are locations completely sheltered from the wind. There are also gusts or heavy turbulence. These are the result of flows that are caused at the interface of air zones of different pressures. On the windward side of an obstacle, there is overpressure which increases with height. This causes a descending flow along the front side, which forms a vortex when it reaches the ground and sweeps around the windward corners. It is considerably increased if there is a small building to windward. The **Venturi effect** is produced by two separate buildings whose axes make an acute or right angle. The pressure of the airflow is concentrated on the gap between the buildings, giving great velocities. **Channelling** is caused when there are urban 'canyons' which concentrate all airflow in one direction.

Fig 11
Wind flow patterns around an isolated building

Global climate change

There is evidence to suggest that from 20 000 years before present (bp) to 11 500 bp global temperatures rose by about 8 °C. They then remained stable until the late 20th century when there was a sudden increase.

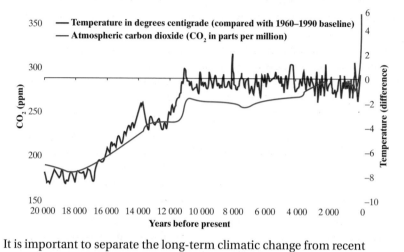

Fig 12
Global temperatures and atmospheric carbon over the last 20 000 years

It is important to separate the long-term climatic change from recent **global warming.**

Evidence for climate change over the last 20 000 years

Changes in past climate are inferred from changes in climatic indicators. These include:

- **Pollen analysis:** Many plant species can only exist in a very narrow range of temperatures and precipitation. This can be further narrowed down if several species of plant are found to have existed at the same time. Pollen from plants is used to identify the mix of plants at any one time. Pollen is unique and is resistant to decay. Once the pollen has been identified, it can be dated using the carbon-14 radioisotope method.

- **Dendrochronology:** The annual growth rings of trees can be used; when climatic conditions limit plant growth, the rings are spaced close together. Counting the rings can give us a date.

- **Ice core analysis:** Cores have been drilled from areas like the Greenland ice cap. Air, trapped in the ice, can be analyzed for carbon dioxide content. The greater the amount of carbon dioxide, the warmer the climate.

- **Historical evidence:** By its very nature, this does not go very far back in time. The Cave of Swimmers, in the Sahara desert of southwest Egypt, contains rock painting images of people swimming, estimated to have been created 10 000 years ago at the end of the most recent ice age. More recently, climate can be inferred, e.g. from farm records of ancient Rome or paintings of London's medieval Thames frost fairs.

Evidence for more recent temperature changes has been gathered from measured data and from satellite data. It suggests that global temperatures are about 0.5 °C higher than they were in 1900.

Possible causes of recent global warming

1. Increasing concentrations of greenhouse gases: The greenhouse effect is a natural phenomenon. Gases like carbon dioxide and methane act as filters in the atmosphere. They allow in all of the shortwave ultraviolet solar radiation. This heats up the earth's surface. This surface emits in return much longer wavelength infrared radiation. The greenhouse gases absorb some of this and the atmosphere heats up. Since the start of the industrial revolution in the 18th century, carbon dioxide emissions have increased. In the late 20th century these emissions accelerated with the industrialization of many less developed countries. This has been added to by an increase in atmospheric methane, a by-product of both the growing cattle-rearing industry and the decomposition of domestic and industrial waste.

2. The removal of huge areas of forest: Trees take in carbon dioxide and act as a **carbon 'sink'**. If the trees are burnt then not only is carbon released back into the atmosphere, but they are often replaced with cattle, who themselves release methane. The enhanced greenhouse effect could have consequences on global, regional and national scales.

Global consequences: Global temperatures are predicted to continue rising. This could lead to melting of the Greenland and Antarctic ice caps; the thermal expansion of oceanic water could cause a predicted global sea level rise of nearly 50 cm by the 2080s. The number of people being flooded each year could increase from 13 million to 94 million. Most of this flooding will occur in southern and southeast Asia.

Regional consequences: Since the early 1970s, the mean annual rainfall has decreased by more than 30% in the Sahel zone of west Africa. This has led to the decrease in available water. There is potential for international conflict over water. The Intergovernmental Panel on Climate Change (IPCC) 2007 report estimated that warming global temperatures are likely to reduce west African agricultural production by up to 50% by the year 2020.

National consequences (the British Isles): Over the next 100 years, the average annual temperature in the south of the British Isles is predicted to rise by around 3.5–4 °C. In the far north, the increase is predicted to be around 2.5 °C. The warming will be greater in the summer. Overall, a slight reduction is predicted in the annual rainfall total over the UK. However, this masks very large seasonal changes. In summer, some parts of the UK will be as much as 50% drier, while in winter they may become up to 30% wetter.

Essential notes

The specification states that candidates must know how groups and individuals have responded to the threat of global warming. Research responses to the threat at an international, national and local level.

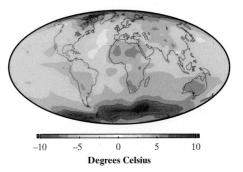

−10   −5   0   5   10
Degrees Celsius

Fig 13
Predicted increase in surface air temperatures, 1960–2060

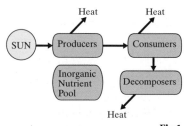

Fig 1
Generalized energy flow through an ecosystem

Nature of ecosystems

An **ecosystem** is a system that includes all living organisms (**biotic factors**) in an area, as well as its physical environment (**abiotic factors**), functioning together as a unit.

An ecosystem is made up of soil, microorganisms, rocks, minerals, water sources and the local atmosphere, each interacting with the others.

Energy flows

All the energy used by living things comes ultimately from the sun. Energy enters living systems as a result of **photosynthesis** by plants and some bacteria. Less than 4% of the incoming sunlight is captured; of this more than half of the energy captured by plants is used in respiration and is lost as heat.

There are two types of organisms that have direct access to the energy in plant tissues:

- **Herbivores** (or primary consumers), e.g. rabbits, which feed on the plant while it is alive.
- **Decomposers**, e.g. fungi and bacteria, which feed on the plant after it is dead.

In most ecosystems, the majority of the energy goes to the decomposers. For example, in an area of grassland only 10% of the energy in plants is taken by grazing animals (e.g. antelope). Herbivores use almost all of their energy intake on respiration and maintaining their bodies; the rest goes to herbivore **biomass** (the flesh and blood of the animal).

Much of the energy in herbivore biomass is taken by:

- **Carnivores** (or secondary consumers), e.g. lions. These meat eaters survive mainly by eating herbivores.
- **Decomposers**.

Almost all of the energy taken in by carnivores goes to maintaining their life systems. The decomposers, which receive most of the plant energy, use up over half of this energy in their life maintenance. The rest may be locked up in soil organic material or taken by organisms that feed on decomposers. Ultimately, all of the energy originally captured by plants is transformed and lost as heat; energy is not recycled.

Nutrient cycling

Nutrients are the chemical elements and compounds needed for organisms to grow and function. Energy flow and **nutrient cycling** are interdependent. The rate of nutrient cycling may affect the rate that energy can be trapped. Plants cannot grow – i.e. make new cells – if essential nutrients are absent.

Nutrients are stored in three compartments within an ecosystem. They are:

- **Soil:** a mixture of weathered rock, air, water and decomposed organic matter on the surface of the earth
- **Litter:** the amount of dead organic matter on top of soil
- **Biomass:** the total of plant and animal life in an ecosystem

This can be summarized by a **Gersmehl diagram** (**fig 2**). When applied to an ecosystem, the size of the circles is in proportion to the amount of nutrients they store.

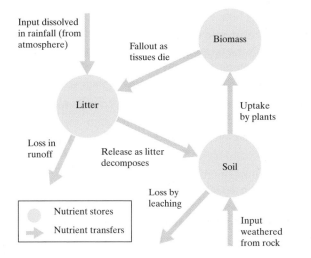

Essential notes

Photosynthesis is a process that turns carbon dioxide into organic compounds, especially sugars, using sunlight. Many kinds of plants, algae and bacteria use it to get food.

Fig 2
A Gersmehl diagram modelling the mineral nutrient cycle

Food chains

The source of all food is the activity of **autotrophs,** mainly through photosynthesis by plants.

- They are called **producers** because only they can manufacture food from inorganic raw materials.
- This food feeds herbivores, called primary consumers.
- Carnivores that feed on herbivores are called secondary consumers.
- Carnivores that feed on other carnivores are **tertiary** (or higher) **consumers**.

Such a path of food consumption is called a **food chain**. Each level of consumption in a food chain is called a **trophic level**.

Examples of simple food chains include:

- Phytoplankton → zooplankton → fish → seal → killer whale
- Grass → grasshopper → frog → heron

Most food chains are interconnected. Animals typically consume a varied diet and, in turn, serve as food for a variety of other creatures that prey on them. These interconnections create **food webs**.

Examiners' notes

Research and draw mineral nutrient cycle diagrams for:

- A temperate **deciduous** woodland
- Heather moorland
- Your chosen **biome** (see below, p. 36–39)

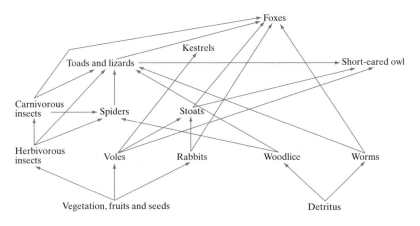

Fig 3
A simplified food web from the acid heathland at Studland, Dorset, UK

Ecosystems in the British Isles over time

Succession and climatic climax

Plant succession is the term used to signify the changes in the composition of a community of plants over time. It refers to the sequence of communities which replace one another in a given area.

A primary succession, or **prisere**, develops by the gradual **colonization** of a lifeless **abiotic** surface. A succession goes through several stages; the entire sequence of stages is called a **sere** and the stages are referred to as **seral stages**.

Succession is a natural increase in the complexity of the structure and species composition (**species diversity**) of a community over time. Plants invade an area when the conditions are suitable and then die off when the succession leads to unfavourable local conditions. This means that there is a change in the dominant plant species with time.

Example of a succession: lithosere

The stages are as follows:

- **Bare rock surface:** This is initially colonized by bacteria that are able to survive on few nutrients and get most of their energy from the sun. The surface conditions are often dry and the soil little more than particles of weathered rock.
- **Seral stage 1 – colonization:** The first plant species to colonize an area are called pioneers. These are lichens that are adapted to the severe (dry, windy, soil-free) conditions. As they die they add dead organic matter to weathered rock and windblown dust. This creates a simple soil which improves water retention. Mosses are then able to develop.
- **Seral stage 2 – establishment:** As the soil develops further, ferns and small herbaceous plants and grasses begin to grow. Species diversity increases. There are more invertebrates living in the soil increasing the organic content. This enables the soil to hold more water.
- **Seral stage 3 – competition:** Larger plants begin to establish themselves. These include shrubs (e.g. gorse) and small trees. They use up a lot of available water and shade the ground. Some of the earlier colonizers are unable to compete and die out.
- **Seral stage 4 – stabilization:** Fewer new species colonize. Complex food webs develop. This stage is dominated by larger, fast-growing trees such as birch and rowan.
- **Seral climax:** This is the final seral stage. It represents the maximum possible development that a community can reach under the prevailing climatic (temperature, light and rainfall) conditions. This is called a **climatic climax** community. In the case of southern England, it is a broad-leaved **deciduous** forest, e.g. oak and ash.

Climatic climax vegetation: the temperate deciduous biome

The plant succession in the UK reached its **climatic climax** around about 8000 years ago. At 5000 years ago the effect of forest clearance by the Bronze Age people of Britain had an effect of creating a **plagioclimax**.

Pioneer weeds — Grasses — Small shrubs — Soft hardwoods (poplar, willow) — Expanding saplings — Mature hardwoods (oak, ash)

Essential notes

Biome: a global-scale ecosystem that shares common climatic conditions. It is a naturally occurring organic community of plants and animals in the climatic climax stage of succession.

Fig 4
General plant succession from bare soil to temperate deciduous forest

In Britain, one typical deciduous temperate forest is oak woodland. This is characterized by four layers which each contain different species:

Tall tree canopy: This comprises oak (the tallest at 30–40 m), beech, sycamore and ash. These trees shed their leaves in winter to reduce transpiration at a time when water is less available. Some woodland is so dominated by a dense, tall tree canopy that light is prevented from reaching lower layers. This reduces the number of species in these lower layers.

Shrub layer: Here there are small trees such as hazel, hawthorn, buckthorn and rowan.

Field layer: This is dominated by woodland flowers such as bluebell, primrose and wood anemone, woodland grasses and sedges.

Ground layer: Here there are mosses, liverworts and tree seedlings. Fallen and decaying wood and leaf litter are also important for fungi.

These different layers may not be obvious in all woods. The composition and structure of woodlands also varies according to age, soil type, climate and landform.

On upland areas with acid soil, woods are dominated by sessile oak and downy birch along with wood sorrel, foxglove and bluebell.

The effects of human activity on succession

Human activity, such as agriculture, regular use of fire or **deforestation**, will lead to a deflected or **plagioclimax** community such as pasture, heather moorland or plantations. This reduces the total number of species present (**biodiversity**). If this land is abandoned, then the plagioclimax community will develop into the climatic climax. These **secondary successions** reach the seral climax much more quickly than primary successions.

Heather moorland is a plagioclimax. It has been created by forest clearance in upland areas. This reduced the soil fertility and allowed hardy plants like heather to dominate. Sheep grazing and/or controlled burning has maintained the moorland and prevented secondary succession.

Examiners' notes

The specification suggests that you use heather moorland as an example of a plagioclimax in the British Isles. Research:

- The heather moorland nutrient cycle
- The reasons for, and the effects of, burning of heather moorland

The biome of one tropical region: the tropical equatorial rainforest

The main characteristics of the equatorial rainforest

Distribution: Equatorial forests are distributed between the latitudes 10° N and S of the equator. They are found in:

- The Amazon river basin of South America, with more than half of this in Brazil; this area holds about one third of the world's remaining equatorial forests
- The equatorial part of Africa including large parts of the Republic of Congo, the Central African Republic and Gabon
- The Guinea coast of Africa in isolated pockets stretching from Liberia to Cameroon
- SE Asia and Oceania, e.g. Malaysia, Indonesia and Papua New Guinea

Manaus (Brazil) 3°S

altitude 44m
annual temperature range 2°C
annual precipitation 2104mm

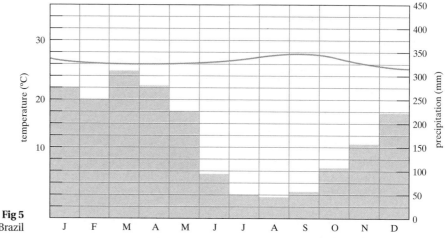

Fig 5
Climate graph for Manaus, Brazil

The climate: It is hot and wet all year round. The average temperatures rarely go below 26 °C or rise above 28 °C. There is a low diurnal range of temperature; during the daytime it might reach above 30 °C, but the nights remain very warm.

Rainforests lie in the **intertropical convergence zone** (**ITCZ**), a low pressure zone where intense solar energy produces a convection zone of rising air that loses its moisture through frequent rainstorms. Rainforests are subject to heavy rainfall, at least 2000 mm. In equatorial regions, rainfall may be year round without apparent 'wet' or 'dry' seasons, although forests on the northern and southern fringes of the region do have a short dry season. During the parts of the year when less rain falls, the constant cloud cover is enough to keep the humidity high and prevent plants from drying out. The climate allows most rainforest trees to be evergreen, constantly shedding and replacing leaves all year round. The moisture in the rainforest from rainfall, constant cloud cover, and **transpiration** creates intense local humidity. Large rainforests (and their humidity) contribute to the formation of rain clouds, and generate as much as 75% of their own rain.

Fig 6 shows that the precipitation in Yaoundé exceeds the potential evaporation for two periods in the year. It is at these times that the ITCZ passes overhead, producing heavy convectional rainfall. This excess water is utilized by the forest for growth.

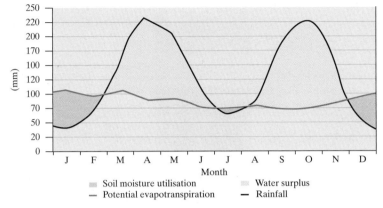

Essential notes

The soil moisture budget is the balance between the incoming precipitation and the outgoing **evapotranspiration**.

Examiners' notes

A detailed case study of one small part of one of the forest areas, with some named species, will help illustrate an essay.

Fig 6
Soil moisture budget for Yaoundé, Cameroon

Vegetation: Rainforests are very productive. The net primary productivity is $2\,200\,\text{gm}^{-2}\,\text{yr}^{-1}$. This means that for every square metre of land, the forest produces 2 220 g of living matter. This is high; it is because it is warm enough for plant growth all year round.

Rainforests are characterized by a unique vegetative structure consisting of several layers. The canopy refers to the dense ceiling of leaves and tree branches formed by closely spaced crowns of forest trees. The canopy is 20m above the forest floor and is penetrated by scattered **emergent** straight and branchless trees that can reach above 50m. Below the canopy ceiling are multiple leaf and branch levels known collectively as the understorey. The lowest part of the understorey, 1.5–6 m above the floor, is known as the shrub layer, made up of shrubby plants and tree saplings. The canopy screens light from the forest floor so that less than 5% of the incident sunlight reaches it; this makes the ground layer quite open. There are few dead leaves and occasional saplings that grow quickly where there is some light.

It is estimated that 90% of the animal and bird species that exist in the equatorial forest biome reside in the canopy. Since the tropical rainforests are estimated to hold 50% of the planet's species, the canopy of rainforests worldwide may hold 45% of life on earth.

Fig 7 shows how the vast majority of nutrients are held in the biomass (vegetation). Those in the litter and soil are rapidly recycled.

Nutrient cycle: The rainforest nutrient cycle is rapid. The hot, damp conditions on the forest floor allow for the rapid decomposition of dead plant material. This provides plentiful nutrients that are easily absorbed by plant roots. However, as these nutrients are in high demand from the rainforest's many fast-growing plants, they do not remain in the soil for long and stay close to the surface of the soil. If vegetation is removed, the soils quickly become infertile and vulnerable to erosion.

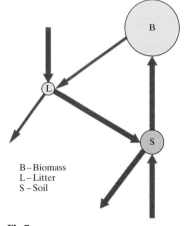

B – Biomass
L – Litter
S – Soil

Fig 7
The mineral nutrient cycle for an equatorial rainforest

☞ This topic continues on the next two pages

Fig 8
The profile of a latosol, a typical rainforest soil

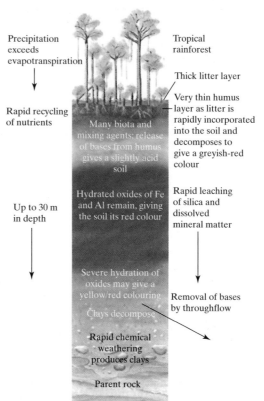

Precipitation exceeds evapotranspiration

Tropical rainforest

Thick litter layer

Rapid recycling of nutrients

Very thin humus layer as litter is rapidly incorporated into the soil and decomposes to give a greyish-red colour

Many biota and mixing agents: release of bases from humus gives a slightly acid soil

Up to 30 m in depth

Hydrated oxides of Fe and Al remain, giving the soil its red colour

Rapid leaching of silica and dissolved mineral matter

Severe hydration of oxides may give a yellow/red colouring

Clays decompose

Removal of bases by throughflow

Rapid chemical weathering produces clays

Parent rock

Soils: Rainforest soils are very deep. The hot wet conditions are ideal for chemical weathering of the underlying rock. This weathering provides minerals for the soil. The surplus water moves down through the soil removing silica-based minerals through the process of leaching. This leaves iron and aluminium-rich compounds that stain the soil red. The leaching leaves the soil nutrient poor (see **fig 7** p. 37).

Ecological responses to the climate and the soil moisture budget
Adaptations by plants:

- The trees grow upwards rapidly to reach the light. This gives them slender, straight, branchless trunks.
- When the crowns of the trees reach the canopy, they spread outwards to obtain as much light as possible.
- Although the soils are deep, the tree nutrients are only found in the uppermost layer. Tree roots spread out laterally rather than penetrate deep into the ground.
- The trees are so high and the roots so shallow that they could be unstable; buttress roots have evolved that stabilize the trees.
- The leaves have developed a waxy cuticle and drip-tips to shed the rainfall. This prevents mould developing.
- **Epiphytes** also grow on canopy trees. These are not parasitic because they draw no nutrients away from the host, but use the host tree only for support.

Human activity and its effect on the biome of the Brazilian rainforest

The main impact of human activity comes from forest clearance. Between May 2000 and August 2005, Brazil lost more than 132 000 km² of forest – an area larger than Greece.

In many tropical countries, the majority of deforestation results from the actions of poor subsistence cultivators. In Brazil a large portion of deforestation can also be attributed to land clearance for pasture or by commercial exploitation of forest resources.

A small number of large landowners clear vast sections of the Amazon for pasture land for cattle; this is sometimes planted with African savanna **grasses** for cattle feeding.

Favourable taxation policies, combined with government-subsidized agriculture and colonization programmes, have encouraged deforestation. The practice of low taxes on income derived from agriculture, and tax rates that favour pasture over forest make it profitable to convert natural forest for cattle ranching.

Soya has become one of the most important contributors to deforestation in the Brazilian Amazon. Due to a new variety of soya bean that flourishes in the rainforest climate, Brazil is on the verge of overtaking the USA as the world's leading exporter of soya. High prices have served as an impetus to expanding soya cultivation.

Soya farms directly cause some forest clearing, but they have a much greater impact on deforestation by occupying already cleared land, thereby pushing ranchers and subsistence farmers ever deeper into the forest. Soya farming also provides a key economic and political impetus for new highways and infrastructure projects, which accelerate deforestation by others.

Fires: Virtually all forest clearing, by small farmer and plantation owner alike, is done by fire. Though these fires are intended to burn only limited areas, they frequently escape agricultural plots and pastures and char pristine rainforest, especially in dry years like 2005.

Examiners' notes

The specification states that you should know what the potential for sustainability is for the equatorial forests.

Make notes on at least one example of a rainforest sustainability project.
Include:

- The location
- The individuals and/or groups involved
- The aims and objectives of the project
- What has been done
- An evaluation of the project

Fig 9
These two NASA Landsat images, acquired in 1992 and 2006, show the growth of the island state of Mato Grosso in central Brazil. In these false-colour images, red indicates vegetation; the brighter the red, the denser the vegetation. The grey-beige colouring in the 2006 photo indicates widespread forest clearing.

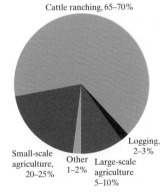

Cattle ranching, 65–70%

Logging, 2–3%

Small-scale agriculture, 20–25%

Other 1–2%

Large-scale agriculture 5–10%

Fig 10
Causes of Brazilian deforestation, 2000 to 2005

Ecosystem issues on a local scale: impact of human activity

Urbanization

Interactions between abiotic factors, such as sunlight and water, and biotic factors, such as plants and microbes, take place in all environments, including urban areas. Unlike natural ecosystems, however, urban ecosystems are a hybrid of natural and manmade elements whose interactions are affected not only by the natural environment, but also by culture, personal behaviour, politics, economics and social organization.

Urban growth alters soil drainage, water flow, light availability and climate. For example, impermeable pavements and rooftops can change an area's hydrology by increasing storm water runoff and can contribute to higher urban temperatures by storing heat energy during the day and releasing it at night.

Urban areas contain a wide variety of habitats from urban forests (e.g. the Mersey Forest) to playing fields; from formal municipal flowerbeds to derelict land. The influence of human activity can make these **habitats** subject to change or can conserve them for the future.

Urban niches: A **niche** is a term for the position of a species within an ecosystem, describing both the range of conditions necessary for persistence of the species, and its ecological role in the ecosystem. In simpler terms, it is the part of an ecosystem that a plant or animal occupies (**microhabitat**).

Colonization of wasteland: An area of wasteland, such as in **fig 11**, contains a variety of niches that are available for plants and animals. These include:

- Horizontal bare tarmac
- Pools of water and damp patches on the ground
- Vertical brick walls
- Open brickwork wall ends
- Piles of rubble
- Wall tops
- Accessible (through broken windows) empty buildings
- Sheltered corners

Fig 11
A variety of urban niches on a small area of wasteland

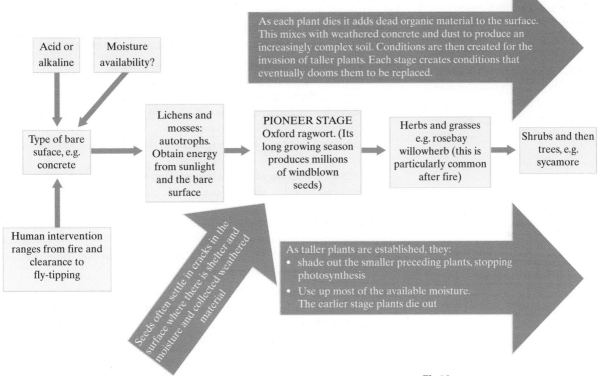

Acid or alkaline **Moisture availability?**

As each plant dies it adds dead organic material to the surface. This mixes with weathered concrete and dust to produce an increasingly complex soil. Conditions are then created for the invasion of taller plants. Each stage creates conditions that eventually dooms them to be replaced.

Type of bare suface, e.g. concrete → **Lichens and mosses: autotrophs. Obtain energy from sunlight and the bare surface** → **PIONEER STAGE Oxford ragwort. (Its long growing season produces millions of windblown seeds)** → **Herbs and grasses e.g. rosebay willowherb (this is particularly common after fire)** → **Shrubs and then trees, e.g. sycamore**

Human intervention ranges from fire and clearance to fly-tipping

Seeds often settle in cracks in the surface where there is shelter and moisture and collected weathered material

As taller plants are established, they:
- shade out the smaller preceding plants, stopping photosynthesis
- Use up most of the available moisture. The earlier stage plants die out

Fig 12
Vegetation succession on urban wasteland

Routeway ecologies

Routeways are distinctive because they allow the incursion of exotic species of plant and insects, brought in by road traffic and trains. Urban routeways act as wildlife corridors, comparable with rural hedgerows.

Railway lines enable animals to move around the city with little or no interference from traffic. During the days of steam trains there were frequent fires which burnt off tall species of plant and allowed the light in, encouraging light-demanding species, e.g. primroses and foxgloves, to establish. Wind-borne seeds are sucked along by the trains, e.g. Oxford ragwort. Spiders are moved along the line in the same way. Also, lack of human disturbance created by the fencing enables foxes and badgers to exist. On land that has not been burnt brambles have established and these provide nesting sites for a wide variety of bird life.

Road traffic acts in the same way with regard to the distribution of animals and insects. Roads also provide food for kestrels and scavenging birds (e.g. magpie). The nitrogen-rich fumes boost the growth of some wildflowers and they in turn increase insects and animals further up the food chain. Many embankments and cuttings are well managed. There has been planned planting of trees and shrubs to act as **noise screens**. Grass is also mown regularly. This can reduce the number of wildflowers and fauna.

Examiners' notes

Questions on routeways will concentrate on the distinctiveness of routeway **ecology**. You have to be able to say what is different about routeways compared to other urban habitats.

☞ This topic continues on the next two pages

Unplanned introduction of new species

A good example of an unplanned introduction, and its potential impact on ecosystems, is the Japanese knotweed. This spreads easily via rhizomes and cut stems or crowns. It is now listed, under the Wildlife and Countryside Act 1981, as a plant that is not to be planted or otherwise introduced into the wild.

Specific problems caused by Japanese knotweed are:
- Damage to paving and tarmac areas
- Damage to archaeological sites
- Reduction of biodiversity by out-shading native vegetation
- Restriction of access to river banks
- Increased flood risk by damaging flood defence structures and through dead stems blocking streams
- Increased erosion when the bare ground is exposed in the winter
- Reduced visibility and access on roads and paths
- Reduction in land values
- Accumulation of litter in well-established stands
- Unsightly appearance
- Cost of treatment (£1 per m^2 for a spraying regime over three years not including re-landscaping)

Changes in the rural/urban fringe

This is the transition area immediately surrounding towns and cities, which is subject to direct urban influences:

Negative: Urban sprawl, land use conflicts and environmental problems are evident in areas of rural/urban fringe around some conurbations. It often accommodates essential but un-neighbourly functions such as waste disposal and sewage treatment, and contains areas of derelict, vacant and underused land as well as agricultural land and woodland suffering from a range of urban pressures.

Positive: The rural/urban fringe can include some special assets, including distinctive landscapes such as river valleys and areas rich in biodiversity, scheduled ancient monuments, listed buildings, reservoirs and canals. These are all of value to communities and contribute to quality of life.

Opportunities and challenges:
- There is a huge demand for recreation in the rural/urban fringe, with more than half of all visits to the 'countryside' taking place within five miles of home.
- 267 country parks absorb 73 million visits per year. They can be improved so that they contribute to wider rural/urban fringe amenity and sustainability needs and enrich landscape character.
- Agriculture in the rural/urban fringe has to deal with all of the problems currently facing agriculture, plus problems such as crime, fly-tipping and land fragmentation.
- High levels of urban deprivation can be found on the urban edge. Involving these communities in the management of the area around them offers community renewal, social inclusion and health benefits.

Ecological conservation areas

Ecological conservation areas differ from abandoned land because succession has been influenced deliberately by humans. Management techniques include:

- The reduction in acidity of old industrial and coal spoil sites by the addition of lime, etc.
- The deliberate clearing of areas to create a variety of habitats for smaller light-demanding species
- Mowing only once a year after meadow wildflowers have flowered

Case study: Chilworth, Southampton

This conservation area is managed by Chilworth Conservation Limited (CCL). The purpose of this newly formed company is to retain the educational and recreational functions of the site with minimal disturbance to its unique habitats and rare species.

Chilworth is on the northern fringe of Southampton close to the M3/M27 junction. It is a small part of what was once a country estate. The area consists of three distinct sections:

- Buxey Wood is ancient woodland with a rich biodiversity. There are distinctive areas of birch-oak wood, hazel-oak wood and alder wood. They each have their own native plants such as bluebells, wood anemone, etc. There are some outstanding mature specimens of beech and yew.

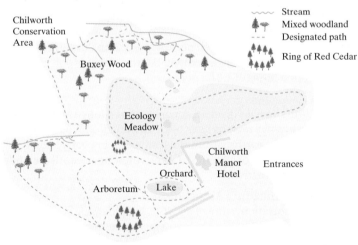

- The Ecology Meadow was once grazed by cattle but is now mown each year to maintain low fertility. The meadow has a large, restored pond that will provide a habitat for amphibians.
- The Arboretum, dating from 1900, contains trees and shrubs from around the world. It also has an orchard and fragments of heathland. The orchard used to be grazed by sheep in late summer, but now it is mown after the orchids have finished flowering.

Fig 13
Chilworth, an example of a conservation area

Essential notes

Recent surveys confirm the importance of Chilworth Conservation Area as a home to:

- 96 species of flowers identified in Buxey West (ancient woodland) and 76 in Buxey East (planted but long established)
- 88 species of fungi
- 55 species of liverworts and mosses
- 3 species of bat: pipistrelles, long-eared bats and noctules

Ecosystem issues on a global scale

Some environmental groups are warning that if current global consumption levels continue it could result in a large-scale **ecosystem** collapse by 2050. They say that the natural world is being degraded at a rate unprecedented in human history.

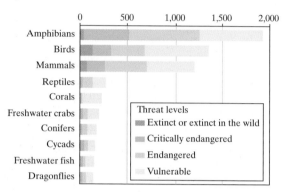

Fig 14
Species under threat

The **ecological footprint** is a measure of human demand on the earth's ecosystems. It compares human demand with the earth's ecological capacity to regenerate. It represents the amount of biologically productive land and sea area needed to regenerate the resources a human population consumes and to absorb and render harmless the corresponding waste.

WWF, a leading conservation body, has concluded that the global footprint exceeded the earth's biocapacity by 25% in 2003, which meant that the earth could no longer keep up with the demands being placed upon it.

WWF argues that, on average, each person needs 2.2 ha to support the demands they place on the environment, but the planet is only able to meet consumption levels of 1.8 ha per person.

Biodiversity

Much of the earth's species diversity is concentrated into a few relatively small areas. Twenty-five regions have been identified, which together cover only 1.4% of the earth's land surface but contain nearly half of all plant species and a third of all terrestrial vertebrate species. All are under pressure from human activities.

Case study: the Amazonia Association

The Amazonia Association, a Brazilian NGO, is developing sustainable livelihoods within the 178 000 ha Xixuau Xiparina reserve. It is located 500 km northwest of Manaus. It supports only 22 indigenous families. The planned expansion of the protected area to 1.5 million ha along the Rio Jauaperi will make it one of the largest conservation projects in northern Brazil.

Many animals that are endangered across the region are frequently seen in the reserve; a recent study recorded 43 mammal species within the area. The Amazonia Association's first action was the establishment of an informal protected area on the Xixuau and Xiparina rivers by pooling land rights among the local inhabitants and incorporating them into the Association.

Essential notes

WWF is an example of a pressure group which believes that the natural world is under threat. There are other groups (e.g. the Lavoisier Group: www.lavoisier.com.au) which take an opposing view. This is an example of contrasting attitudes to an issue.

The reserve is known in Brazil as an **extractivist** area. This is an area used by traditional communities based on small-scale extractive activities (such as harvesting rubber and Brazil nuts) and subsistence agriculture.

When indigenous people leave the rainforest, the natural environment is destroyed as commercial, large-scale fishing fleets, loggers, hunters and ranchers move in. In contrast, the Xixuau Xiparina reserve is a model of conservation and small-scale sustainable economic activity, acting as a magnet for the region. This has drawn a small number of families who have returned to their traditional lands, committed to preservation of the forest.

Exploitation of the natural resources is forbidden; logging is only for use by the community. Fishing and small-scale farming provide basic food stuffs, topped up by income from activities such as ecotourism and nature documentaries. Three primary schools are operating, a nurse is in training, the health post is well stocked and malaria has practically been eradicated in the area. Communication with the outside world is via a satellite telephone and internet powered by solar panels. Future planned projects include a secondary school, a forest college, a research centre and diversification of economic activities, e.g. beekeeping.

Following a long campaign by the local communities, the reserve is about to be formally protected and increased to around 600000 ha. The Amazon Charitable Trust aims to help make it the first profitable extractivist reserve in Roraima state. It is hoped that it will link together with other areas to form the Central Amazonian Ecological Corridor. The aim of this is to protect a continuous area of high ecological importance reaching across the Brazilian Amazon from the border with Peru to Guyana.

Action for an expanded reserve

The Amazonia Association believes that while only the local people are in a position to offer protection to the forest, the support of government agencies is vital to ensure a permanent solution to the threats that the area faces. They are actively working to strengthen the protection of the Xixuau Xiparina reserve, and expand it as a legally defined conservation area: the Lower Rio Branco-Jauaperi Extractivist Reserve (Resex). The Resex will be a conservation unit that ensures permanent rights of residency to the families living in the forest, guaranteeing them the right to manage natural resources in carefully defined, sustainable ways.

The Resex will involve 10 traditional communities spread over an area of 600 000 ha along both sides of Jauaperi river, in the states of Amazonas and Roraima. It will include the existing Xixuau Xiparina reserve. The process of designating the Resex is being carried out in partnership with the Brazilian Ministry of the Environment. Its creation has been approved at every level of Brazilian government and now simply awaits a presidential decree in order to come into effect. The campaign to accelerate its signature by the Brazilian president has received support from several local organizations.

The expansion of both the protected area and the number of communities included in it represents a big challenge for the Amazonia Association and local partners: more work, more human resources and more income will be needed in order to extend to the whole region the same standard of living achieved by the inhabitants of the Xixuau community.

Examiners' notes

The specification states that you should have **two** contrasting management schemes in fragile environments. Choose another area and:

- Locate it
- Describe the fragility of the environment
- Explain how the management plan overcomes the desire to overexploit the area
- Describe what issues are involved

Essential notes

Some groups and individuals are opposed to the establishment of reserves like Xixuau Xiparina. Research their motives and any actions they have taken to further their cause.

Urbanization

A high proportion of the world's population lives in large urban areas. There were 83 cities with populations of more than 1 million in 1950, 34 of them in developing countries. By 2000, there were over 280 such cities and this number is expected to have doubled by 2015. All the new **millionaire cities** (cities with populations over 1 million), and 12 of the world's 15 largest cities, are in developing countries. Nineteen **megacities** of 10 million people existed in 2007, with Tokyo, the largest, having 35.2 million inhabitants. In the span of half a century several cities have more than tripled their population.

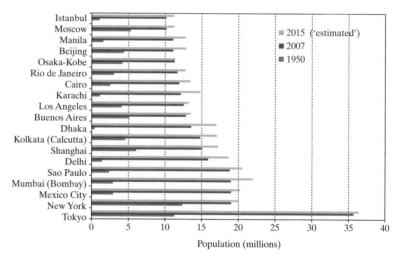

Fig 1
Population of the world's megacities, projected to 2015

Consider how you would describe and comment on the data in **fig 1**. What changes have occurred? Can you identify particular trends? Why might these changes have occurred?

Contemporary urbanization processes

These include: urbanization, suburbanization, counter-urbanization and re-urbanization.

Urbanization is growth in the proportion of a country's population that lives in towns/cities as opposed to rural areas.

Case study: Delhi

According to the 2001 census of India, the population of Delhi was 13 782 976. In that year alone, the population had increased by 285 000 as a result of migration and by an additional 215 000 as a result of natural increase; this made Delhi one of the fastest growing cities in the world.

Causes of growth

- Delhi is India's capital city. It is the economic, political and administrative hub of the country, attracting migrants in all these sectors. It is also an important light industrial centre, with over 130 000 industrial units, producing everything from TVs to medicines.

- Delhi has also now emerged as the fashion capital of India – more than 60% of the design community resides in this city (one of the best known being Ritu Kumar, famous for her bridal collections).
- Delhi's literacy rate of 81.8% compares favourably with a national rate of 66%, and 57% in neighbouring Uttar Pradesh. Many of India's most respected universities and research institutes are here, such as the University of Delhi and the Jawaharlal Nehru University. There are also five medical colleges and eight engineering colleges.

Problems

- It is estimated that 52% of Delhi's population live in slum conditions. In comparison to Delhi's average infant mortality rate (IMR) of 40, the IMR in slums is higher, at 54 for every 1 000 live births. The immunization level is a dismal 34% and because of the lack of safe drinking water and sanitation, there is a high incidence of diseases such as diarrhoea and anaemia. For 31% of Delhi's slum dwellers there are no sanitation facilities and no underground sewage system.
- It has also been claimed that Delhi has the world's worst air pollution. By 2008, Delhi's average PM_{10} level was 218, the sixth worst in India, and more than four times the level the World Health Organization considers safe. India's Chittaranja National Cancer Research Centre says that one in three Delhi residents has at least one respiratory problem as a result of this.
- The Yamuna is one of the most polluted rivers in the world, especially around Delhi, which dumps about 58% of its waste into the river. Delhi alone contributes 3 296 MLD (million litres per day) of sewage to the river. The problem is aggravated by the fact that the water remains stagnant for almost nine months of the year, until the summer monsoon arrives.

Solutions?

- The government has spent nearly $500m trying to clean up the Yamuna river by repairing and rebuilding the sewage system.
- New towns such as Noida, 20 km SE of Delhi, have helped to alleviate the pressure on the capital.
- Delhi's new metro opened in December 2002 and has been instrumental in helping to relieve air pollution and traffic congestion.
- The city's bus, taxi and autorickshaw fleets have been converted to CNG (compressed natural gas), a cleaner alternative.
- The Salaam Baalak Trust is an NGO providing shelter, education and healthcare for street children.
- Another NGO, ASHA (meaning 'hope' in Hindi), works to improve healthcare in Delhi, especially through the empowerment of women.

Essential notes

Many cities in the developing world are still growing rapidly. There are two reasons for this:

- **Rural–urban migration** as a result of push and pull factors.
- **Natural increase** due to the fact that most migrants are in their reproductive years.

Suburbanization

Suburbanization is the process of population movement (and increasingly industry and retail) from the central areas of cities to the outskirts (the suburbs), often engulfing surrounding villages/rural areas. It has been facilitated by improvements in both public and private transport and has led to the creation of residential estates – often with associated services: schools, GP practices, pubs, parks and playgrounds, etc.

In the UK, housing type typically varies from semi-detached to large, executive detached and bungalows. Such areas are popular with middle to higher income families seeking to escape problems associated with urban life: stress, pollution, congestion, crime. Increasingly, services and industry have moved out too; out-of-town shopping centres such as the Trafford Centre in Manchester can take advantage of cheaper rents, more space, better access, etc. This has also led to planning and management issues to curb urban sprawl; for example, green belts.

Case study: Ballawattleworth estate, Peel (Isle of Man)

Suburbanization may be associated with an increasing urban population; this has certainly been true in the town of Peel, as suburbs such as Ballawattleworth have developed on the outskirts. The Isle of Man census 2006 lists the population of the town as 4 280: an increase of 11.57% from the population of 3,785 in 2001.

There has been a major impact on both the local primary (Peel Clothworkers) and secondary (Queen Elizabeth II High) schools in terms of increasing pupil numbers and class sizes. Both schools initially had to expand into prefabricated accommodation before more permanent provision could be built. In the case of the QEII, this has meant construction of new dining facilities, additional classrooms and a new KS5 centre.

Another positive effect has been the increased public pressure for improved leisure amenities, and a new swimming pool (the Western Swimming Pool) has been erected adjacent to the Ballawattleworth estate. The estate itself has generated its own service provision – Busy Bears Nursery and the Highwayman public house among others. Such facilities can also be utilized by the original residents of the town – a positive benefit for all. The estate has also provided some much-needed housing for first-time buyers, as there is a mix of apartments, two-bedroom semi-detached and executive style homes.

Fig 2
Typical suburban housing available on estates such as Ballawattleworth in Peel

Counter-urbanization

Counter-urbanization is the migration of people from major urban areas to smaller towns, villages or rural areas – often 'leapfrogging' the green belt. This process again reflects people's desire to move away from what are perceived as congested, polluted, crime-ridden urban environments to more pleasant rural environments where land and housing prices are cheaper. As with suburbanization, it has been facilitated by increasing car ownership, improvements in public transport and improvements in technology which allow tele-working, for example through video conferencing facilities and the use of the internet in general.

Case study: Crosby (Isle of Man)

The appearance of this village, 6 km from the island's capital, Douglas, is changing, with increasing infill development along the main A1 commuter route between Douglas and Peel. This has occurred as local builders have seized on the opportunity to build on unused plots in what has become a very desirable location. In addition, newcomers tend to want modern facilities and therefore add conservatories, garages, loft conversions, etc. They like the rural image of village life, but it has to be 'sanitized' with the addition of modern conveniences. An excellent example is the Eyreton Farm barn conversions. These still look like picturesque barns, but the stonework is re-pointed, they are triple glazed, constructed to a very high specification and the grounds around them have been landscaped.

The village pub itself, the Crosby, a stopping-off point along the motorcycle TT course, has undergone a programme of renovation and extension along with a refurbished play park and the addition of a skate park and BMX track. It could certainly be argued that such services have been necessary to serve the needs of the expanding village and perhaps cater for the more 'sophisticated' demands of the newcomers.

Examiners' notes

Again, all the required elements are here: the definition, the characteristics of the process and the effects of that process in a specific area. There is also some reference to planning and management.

Examiners' notes

Suburbanization and counter-urbanization are often confused – be careful!

Re-urbanization

Re-urbanization is the movement of people back into urban areas, particularly the inner city or even the central business district (CBD) itself. It is often associated with **urban regeneration schemes** or **gentrification**. Increasingly, it is also associated with the move towards **sustainable communities**.

Gentrification

This is the process by which older, often rundown housing areas (usually close to the city centre) become desirable once again and are physically and socially upgraded. Professional groups, such as doctors, lawyers and teachers, are attracted by the character of the housing – typically Edwardian and Victorian terraces – and its greater accessibility to the CBD. They move in and renovate the property, usually on an individual basis. As a result, other property in the area becomes more sought-after and the social composition of the area gradually changes.

The purchasing power of the new residents is higher and this leads to an increase or an upgrading in local services such as wine bars, restaurants, delicatessens, boutiques – all attracted by the possibility of catering for the new, wealthier clientele. New residents also tend to lobby for improvements to the area in general, such as traffic-calming measures and the addition of street furniture.

There are many examples of gentrified or even 'super-gentrified' areas within Britain's towns and cities. The latter term refers to areas requiring even higher salaries and bonuses, such as Islington, popularized by former prime minister Tony Blair in the late 1990s. Interestingly, the borough as a whole is still classed as the eighth most deprived local authority area in Britain, which demonstrates another aspect of gentrification: the gap in wealth between the original and newer residents, which may itself give rise to friction. Inner-city schools, for example, fail to see a similar level of improvement as other local services, as the upwardly mobile newcomers often prefer to send their children to private kindergartens and then to fee-paying schools.

Case study: Notting Hill, London W11

In Victorian times, Notting Hill was a rough, working-class area. By the 1950s the area had become synonymous with slum landlords and inner-city deprivation. In 1958, it was the scene of race riots between the Afro-Caribbean community and local 'Teddy boys'. A riot during the infamous Notting Hill Carnival of 1976 inspired the Clash's punk anthem 'White Riot'.

However, over the past 30 years Notting Hill has become synonymous with gentrification, with rocketing property prices and estate agents coining names like 'Hillgate Village'. Period houses on Kensington Park Road typically retail at about £4 million. Notting Hill's secluded communal gardens make it one of London's most desirable areas for wealthy families. *Notting Hill*, the movie, has helped popularize the area, but gentrification was under way long before this.

Trendy eating places include the Fat Badger gastropub, the Notting Hill Brasserie and Feng Sushi. Also famous is Notting Hill's Electric Cinema, reopened in 2001 on the back of the area's increasing affluence.

Urban decline

Characteristics and causes of urban decline

Characteristics of urban decline include:

- High population out-migration figures
- Boarded-up shops and housing
- Empty/derelict properties
- Vacant factories and overgrown wasteland
- Reduction in service provision, e.g. closure of schools
- Low levels of educational attainment
- High levels of crime, vandalism, graffiti
- Political marginalization of residents

Inner cities

Problems facing inner cities include:

- **Economic decline:** this has occurred both with deindustrialization and the movement of lighter, hi-tech and service-based industries to peripheral areas of cities (or indeed to small towns and rural areas). Such areas can provide more space for development and expansion at a cheaper price and with better access than the inner city.
- **Population loss:** this has occurred in tandem with the above as the younger, more mobile, more skilled and more affluent population has also moved out of the UK's inner cities in search of housing and employment in a more pleasant environment. Between 1951 and 1981 the UK's largest conurbations lost 35% of their population.
- **Poor physical environment:** the physical environment of inner cities can also be somewhat unappealing and intimidating.
- **Political problems:** inner cities have some of the lowest election turnouts in the UK, perhaps illustrating the degree to which residents feel neglected. In the 2010 parliamentary elections, the turnout in Hackney was only 61%.

Case study: the decline of the London Docklands

- The Port of London's prosperity began to decline from the 1960s onwards when increasingly large vessels were unable to access the upper docks, such as St Katherine's. Meanwhile, there was intense competition from new deepwater facilities at Felixstowe, Tilbury and Dover. Increasing mechanization and containerization also meant fewer jobs for the dock workers.
- East London itself (broadly including Tower Hamlets, Newham, Southwark, Lewisham and Greenwich) was to lose some 150 000 jobs between 1966 and 1976, which represented 20% of jobs in the area. The hardest-hit sectors included transport, distribution, and food/drink processing, all sectors specifically related to port activity.
- Unemployment was accompanied by population decline. Between 1971 and 1981, the Dockland boroughs lost, on average, 30% of their population. Southwark was hardest hit, with a decline of 38%.
- Poor housing was another feature of the Docklands. By 1981, 80% of people lived in poor quality high-rise council flats or houses. Of these, 30% were classed as being unfit for human habitation.

Essential notes

Nineteen London boroughs fall within the 50 most deprived local authority areas in terms of income, employment, health, education, housing, crime, living environment, or a combination of these factors. In 2007, Hackney was classed as the second most deprived local authority area in England, after Liverpool. Figures published in 2009 placed the borough as fifth for crime rates within London as a whole, with an unemployment rate of 6.7% as opposed to the London average of 4.3% and the national average of 4.1%. A further challenge for the borough is that only 44.5% of school children claim English as a first language.

Examiners' notes

The list of inner-city issues is useful but you will score higher marks in the examination if you can provide supportive detail of your chosen location.

Urban regeneration

Property-led regeneration

This is chiefly associated with **urban development corporations** (UDCs) set up in the 1980s and 1990s, such as the London Docklands Development Corporation (LDDC). UDCs were given planning approval powers over and above those of local authorities, and the focus was on the physical, social and economic regeneration of inner-city areas using public money to attract private investment. However laudable the aims, there have been some criticisms – for example, local people often complained that they had no say in the developments taking place. In the case of the LDDC, many of the new hi-tech jobs were not suitable for the traditional East End dock workers, who lacked the appropriate skills. Likewise, it is arguable that the affluent 'yuppie' newcomers may have pushed house prices beyond the reach of the original residents, who felt resentful that their once tight-knit community had been changed beyond all recognition.

Achievements of the LDDC (1981–1996):

- **Physical regeneration:** Conservation groups have supported the environmental regeneration of the area: 160 000 trees were planted, 17 conservation areas were created and several new parks. The Thames Barrier Park, adjacent to Pontoon Dock, is an 8.9-ha park intended to aid the regeneration of the area by creating an attractive public space alongside residential and commercial developments.
- **Social regeneration:** Almost 8 000 local authority homes were refurbished and owner-occupied homes increased from 5% to 40% in this time period. A post-16 college and a technology college were constructed along with a national indoor sports centre and a marina for watersports. Other developments included the Surrey Quays shopping centre and the ExCeL exhibition centre. In all, 120 000 new jobs were generated and unemployment fell over the period from 14.2% to 7.4%.
- **Economic regeneration:** The national government financed the Isle of Dogs Enterprise Zone, encouraging private investment by providing initial financial leverage. London City Airport was built, connecting to the City of London and the West End via the Docklands Light Railway (DLR). Property developers Olympia and York were responsible for the construction of office blocks, such as the flagship Canary Wharf. These facilities attracted companies such as the Telegraph newspaper group and the New York Stock Exchange.

Partnership schemes

Partnership schemes between local and national government and the private sector are chiefly associated with:
- City Challenge partnerships, e.g. Hulme City Challenge
- Prestige project developments, e.g. Cardiff Bay
- Sustainable communities, e.g. Greenwich Millennium Village

City Challenge partnerships: These were based on a system of competitive bidding by local authorities who had to develop imaginative plans involving the private sector and the local community to gain funding. By the end of 1993, over 30 City Challenge partnerships had been established and by the end of 1997 the government was able to claim that over 40 000 homes had been improved, 3 000 new businesses had been established, 53 000 jobs had been created and nearly 2 000 hectares of derelict land had been reclaimed.

One particularly well-known example is the £37.5 million **Hulme City Challenge** partnership in Manchester, which brought together organizations such as The Guinness Trust, Bellway Homes and Manchester City Council. The crescents and deck-access housing of the 1960s were demolished, as the alleyways had become synonymous with crime, vandalism and drug abuse. These were replaced with traditional two-storey homes and low-rise flats with attractive squares and courtyards. A new community centre, the Zion Arts Centre, provides dance and music facilities for young people, and the main shopping centre has been completely refurbished with the addition of an ASDA supermarket.

Examiners' notes

Hulme would make a good case study. Find out more at: www.cube.org.uk/ftp/City/Tours/cube_tours_hulme.pdf

Fig 3
The Hulme Arch, an architectural showpiece, was erected as a symbol of new hope and local pride

This topic continues on the next two pages

Prestige project developments: These include waterfront developments such as those in Cardiff Bay, the International Convention Centre area of Birmingham and the St Stephen's development in Kingston upon Hull. They are known as 'flagship projects', as they involve the creation of innovative and eye-catching developments which aim to lead the way in regenerating such areas.

Examiners' notes

Alternatively, why not research one of the 'prestige' projects for your case study? The official Cardiff Bay website has a very useful education section at: www.cardiffbay.co.uk

Fig 4
Cardiff Bay

Sustainable communities: These were initiated by the Labour government in the early 2000s. They are defined as places where people will want to live and work both now and in the future. They are intended to meet the diverse needs of existing and future residents, be sensitive to their environment, and contribute to a high quality of life. They should be safe and inclusive, well planned, well built and well run, and offer equality of opportunity and good services for all.

Case study: Greenwich Millennium Village (GMV)

English Partnerships, now part of the Homes and Communities Agency (HCA), had the overall responsibility for this project and invested over £200m. The site covers 121 ha of what was the largest gasworks in Europe and is being developed by a consortium of Countryside Properties and Taylor Wimpey.

The housing is of modern, environmentally friendly design. There is extensive use of glass, split bricks, corrugated panels, timber cladding and zinc sheeting – materials selected for their 'green' credentials. The development aims to cut primary energy use by 80% using low-energy building techniques and renewable energy technologies. There is also a primary school and health centre with timber-clad buildings, which reduce energy consumption by maximizing daylight and by using more efficient systems for heating and air conditioning.

GMV is planned by the developers to continue to expand until about 2015, with its own integrated village shopping and community centres. As of 2008, 1 095 homes and a village square with shops had been completed, with a further 1 843 homes planned.

Essential notes

Note the specific detail given here: the background to the scheme, the partners involved, the achievements (with facts and figures to support the claims) and even a critique. Whichever case study you decide to use, aim for a similar depth and breadth of knowledge.

Examiners' notes
Whenever you are asked to 'evaluate' the success of a particular scheme, it is a good idea to know about some of the problems as well as the successes.

Fig 6
The ecology park at GMV

The importance of a natural environment was recognized by English Partnerships throughout the development at Greenwich peninsula. Three main areas of parkland have been created including an ecology park, and works have been carried out to improve the riverside environment.

Is it all good? Not everyone agrees that GMV is an unqualified success. The only way to get from GMV to Greenwich (the community of which it is supposed to be a part) is by the notoriously infrequent 129 bus. It is cut off from the surrounding traditional Edwardian and Victorian areas by the Blackwall flyover, but it is very well connected to Canary Wharf, where most of its inhabitants work, creating early morning bottlenecks. Few families are to be seen, which has led to the area being nicknamed 'yuppie village', a middle-class young professional enclave set among what many would describe as wasteland, with extremely high rents and flat prices to match.

Essential notes

The specification requires knowledge of both the *causes* and the *impacts* of the decentralization of retailing and other services. As always, it is a good idea to have some examples to back up general points made.

Examiners' notes

There is some similarity here with the reasons for suburbanization (see p. 48) – this should help your learning, but think carefully and clearly how to apply this knowledge in the examination.

Retailing and other services

Trends over the last 30 years:

- The desire for 'one-stop' shopping has precipitated the development of large supermarkets and hypermarkets in town centres and suburbs.
- Large **out-of-town** shopping centres and retail parks, offering a multiplicity of services, have developed on the periphery of large towns and cities.
- The widespread use of the internet is promoting home shopping and home delivery services. British company Tesco was the first retailer in the world to offer a robust home shopping service in 1996.

Why has this happened?

- Increasing public dissatisfaction with the quality and accessibility of traditional CBD shopping areas – congestion/vandalism/difficulty and expense of parking, etc.
- Improvements in transport infrastructure have made out-of-town shopping centres increasingly accessible, e.g. rapid motorway transit with slip-road access (such as the M60 for the Trafford Centre). The increase in private car ownership has also allowed people to travel greater distances with ease; and developments such as the Sheffield Supertram, and frequent subsidized rail services to local towns such as Barnsley and Doncaster, have made Meadowhall accessible to those without cars.
- Land prices are much cheaper out of town. This allows developers room for expansion and the chance to provide many shops and services under one roof at low density, with landscaped surroundings and free parking, attracting many customers. Many such developments make use of brownfield sites (e.g. Meadowhall built on the site of the former Hadley's steelworks) even 'politically' desirable.

Effects on traditional CBDs:

- Loss of retail function, especially food, electrical goods and DIY
  - This can lead to areas of 'discard', characterized by derelict or boarded-up buildings and increasing numbers of low-grade shops, such as discount stores and charity shops.
  - Rundown city centres can appear dangerous, especially at night, further discouraging investment.
- Loss of offices to suburban locations
  - These too are attracted to peripheral areas in search of more prestigious locations in landscaped science or business parks, further compounding the problem.
- Increasing costs of upkeep and development of the CBD itself
  - Although this is a problem, town centres may seek to reinvent themselves through a process known as 'fight-back'. Examples include Sheffield's Heart of the City regeneration scheme.

Case study: Trafford Centre, Manchester

- Highly accessible – located adjacent to junctions 9/10 of the M60 with convenient access to the M602 and an excellent dual carriageway link to the city centre.
- 10 000 free parking spaces. Car users entering the site encounter a vehicle messaging system that communicates messages about car park availability on site.
- The X50 service is a new express bus link to Manchester Piccadilly, offering journey times of just 25 minutes.
- Latest additions include Barton Square, a unique home retailing section including Next Home and Legoland.
- The catchment area is larger and more populous than any other regional shopping centre in the UK, comprises 5.3 million people within a 45-minute drive, with a total potential retail expenditure of £13bn.

The centre boasts:
- 230 stores, including six nationally renowned anchor stores, such as John Lewis and Selfridges
- A 20-screen cinema
- Laser Quest
- An 18-lane bowling alley
- Europe's largest food court with seating for 1 600, and 60 restaurants, cafés and bars providing over 6 000 covers
- Three Premier Inns ring the area, making it an attractive option for mini-breaks.

The Trafford Centre's interior and exterior architecture is mock Rococo/Late Baroque in design, and decorated primarily in shades of white, pink and gold, with ivory, jade and caramel-coloured marble throughout.

Trafford Quays businesses:
- Chill Factor – Manchester's indoor 'real' snow centre is home to the UK's longest indoor real-snow ski slope, at 180 m
- An Airkix indoor skydiving centre
- The PlayGolf driving range, with 64 driving bays
- The Powerleague Soccer Dome, with 19 indoor and four outdoor pitches, organized matches, five-a-side leagues and football camps
- David Lloyd Leisure offers fitness and health facilities

Fig 6
Trafford Centre, Manchester

Examiners' notes

Again, the specification states that you must know about one urban centre that has undergone redevelopment. It would be a good idea (although not compulsory!) to match this with the out-of-town centre you have studied.

Redevelopment

Manchester city centre redevelopment

Although it is difficult to prove conclusively that the developments at the Trafford Centre have had an adverse effect on trade in Manchester's CBD, it is perhaps not surprising that there has been an element of 'fight-back' to remain competitive.

The city centre has its own large indoor all-weather shopping mall, the Arndale Centre, significantly redeveloped after the 1996 IRA bombing. The commercial focal point is the new Marks & Spencer, the largest in Europe. Other flagship stores include Selfridges and Harvey Nichols. There are also many leisure facilities in the city centre, including the Printworks – containing a multi-screen cinema (including an IMAX screen), numerous bars, clubs and restaurants and also Manchester's first Hard Rock Café.

The landscaping of the city centre has provided several public spaces including the newly developed Piccadilly Gardens and Exchange Square, both of which are used for screening public events. Two of the city centre's oldest buildings, The Old Wellington Inn and Sinclair's Oyster Bar, were dismantled, moved 275 m and re-erected in 1999, to create Shambles Square.

Fig 7
Shambles Square, Manchester

Examiners' notes

Some examination questions about redevelopment schemes in urban areas may be couched in very general terms. In this case you can choose whether to write about redevelopment of inner-city areas or CBDs. However, some questions may specify one or the other, so do read the question carefully!

Special events which draw visitors include the Christmas markets, and the annual Spinningfields ice rink. Spinningfields is a large business, retail and residential development that lies in the western portion of the city centre, described as becoming the 'Canary Wharf of the North'.

There is a renowned 'gay village' around the Canal Street area in the east of the city centre, which hosts an annual Gay Pride festival, and a large Chinatown with numerous far eastern-style restaurants.

Meanwhile, the Northern Quarter is now regarded by many as the central district's creative hub. It is home to numerous bars, such as the cocktail bars Apotheca and Trof, and live music venues such as the jazz bar Matt & Phred's.

Contemporary sustainability issues in urban areas

Waste management: recycling and its alternatives

The pyramid in **fig 9** shows the various options with regards to **waste management** – the most favoured option is at the top and the least favoured (but most commonly used!) is at the bottom.

Waste management is a key **sustainability** issue. The average person in the UK produces 517 kg of household waste every year. What are the options?

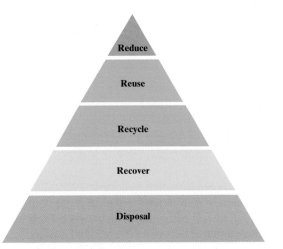

Fig 8
The waste management hierarchy

Examiners' notes

The specification does not require a case study for either waste or transport management. However, an essay question could still be set on either of these topics, so you need to have plenty of examples to refer to, preferably from a range of countries at different stages of development.

Reduction: Businesses can be encouraged to reduce the amount of packaging used and to encourage consumers not to use plastic bags. Tesco's Bag for Life scheme is one example. Designing products that use less material to achieve the same purpose would also reduce waste – for example, the 'lightweighting' of beverage cans.

Reuse: There is some reuse of milk bottles, glass drinks bottles and even jam jars; the 'bags for life' schemes are the most successful.

Recycling: Although there are costs involved here (in the start-up of schemes, transport of materials and the energy needed for reprocessing), this is a practical alternative for paper, glass, cans, plastics and clothes. The energy saving from recycling just one bottle will power a computer for 25 minutes. On a small scale, composting can also be included under recycling – waste materials that are organic in nature, such as paper products, plant material and food scraps, can be recycled as mulch or compost for agricultural or landscaping purposes.

Energy recovery: This is achieved through incineration, and composting of biomass.

- **Incineration:** Modern incinerators can convert waste material into energy, generating electricity or powering neighbourhood heating schemes, so this could be regarded as sustainable.
- **Composting:** On a large scale, anaerobic digestion can be accomplished in enclosed reactors to produce biogas to provide an energy supply.

Essential notes

Try investigating Emma Bridgewater's 'Take an Old Bag Shopping'.

☞ This topic continues on the next two pages

Essential notes

National waste targets for England

- 40% of household waste to be recycled or composted by 2015; 50% by 2020.
- Reduction in the amount of household waste not reused, recycled or composted of 45% by 2020.

Essential notes

Websites worth investigating:

www.tesco.com/greenerliving/at_home/default.page

http://plana.marksandspencer.com/we-are-doing/waste/packaging/

www.sita.co.uk/

www.mariavazphoto.com/curitiba_pages/curitiba_recycling.html

http://greenjobsforindia.blogspot.com/2009/09/dharavis-recycling-potential.html

www.preston.gov.uk/rubbish-waste-and-recycling/household-waste-and-recycling/beacon-status/px

Disposal: In the UK, disposal equates with landfill. Waste is dumped in old quarries or hollows where it is unsightly and often a threat to groundwater supplies and river quality as toxic chemicals are leached out. Decomposing waste also emits methane, the most toxic of the greenhouse gases and potentially explosive.

Transport and its management

This concerns the development of integrated, efficient and sustainable systems. In the UK, the increase in suburbanization and counter-urbanization has been accompanied by growth in disposable incomes and a trend towards multiple car ownership within families. This pattern is repeated throughout the developed world and it seems likely that developing countries will follow suit, with the associated problems of pollution and congestion. How can sustainability be achieved?

Case study: London

The London **congestion charge** aims to reduce congestion, and raise investment funds for London's transport system. Before the scheme was introduced, it was estimated that congestion was costing businesses £2m a week!

The zone was introduced in central London in 2003, and extended into parts of west London in 2007. A payment of £8 is required each day for each vehicle which travels within the zone between 7 a.m. and 6 p.m. (Monday–Friday); a fine of between £60 and £180 is levied for non-payment. A Transport for London report (June 2006) stated that a reduction in congestion of 26% has been achieved. The scheme generated net revenues of £122m in 2005/6 which have largely been spent on improved bus services.

A pleasing background trend has been the reduction in personal road accidents within the zone – estimated at 40%–70% per year.

The combined effect of congestion charging and improved vehicle technology has led to a reduction in NO_x emissions of 13% and a reduction in PM_{10} emissions of 15% since 2002.

Case study: Curitiba

Jaime Lerner was first elected mayor of Curitiba, Brazil, in 1971. With a background in architecture and urban planning, Lerner was at the forefront of innovative transport initiatives that have attracted global interest. He was a leading figure behind the creation of the IPPUC, Curitiba's urban planning and research institute. Their architects and engineers perceive the necessity for an integrated system and believe that this is far easier to achieve through this single organization than if the city had a variety of independent firms and contractors. Curitiba has five structural arteries that run east to west, and development along these has been encouraged, diverting traffic from the city centre and allowing it to become a pedestrian-friendly area.

The bus service has been designed to function like a subway system, transporting large numbers of people along numerous routes, and has proved more reliable and easier to use than a private car. The old, noisy and polluting buses have been replaced with cleaner and more efficient models. These are locally assembled by Volvo, reducing transportation costs for the city that would have been vastly inflated had the buses been imported. Passengers board and alight via a special tube on Curitiba's central transit routes so that boarding is not delayed by fare collection. The glass tube stations provide citizens with a clean, protected area in which to wait for the bus. The platform of the tube station is parallel to that of the buses, so there are no awkward steps to climb and the bus is 'wheelchair accessible'.

The alternative might have been excavation for a subway, but this can take years if not decades, and the money that Curitiba has saved has been allocated to other social causes. For example, in 1976, the city adopted a slum relocation plan to assist low-income families in building low-income housing near the centre of the city. This has created socially integrated neighbourhoods that provide public health, education, day care centres, and recreational services. By meeting the needs of the poorest, the city has also saved money and energy because low-income neighbourhoods, with all the necessary amenities, reduce the need for travel. Sustainability indeed!

Examiners' notes

Again, although not strictly required by the specification, the London and Curitiba case studies provide interesting and comparative detail which would certainly help in the construction of an essay.

Development

In a geographical context, the term 'development' implies an improvement in living conditions. The specification lists the following areas where change may be experienced: economic, demographic, social, political and cultural – so it is important to understand what these mean:

- **Economic development:** This suggests an increase in a country's level of wealth which may well be associated with a change in employment structure.
- **Demographic development:** This would include indicators such as an increase in life expectancy, a reduction in death rates and infant mortality rates along with falling birth rates.
- **Social development:** This relates to improvements in education and literacy, healthcare, housing, sanitation – but also to greater levels of personal freedom and opportunities for advancement.
- **Political development:** This implies that a country's population will have a greater say in the development of the institutions, attitudes, and values that form the political basis of their society. The country therefore becomes more democratic.
- **Cultural development:** This encompasses empowerment/equality of women, and improved race relations in multicultural societies.

There are many ways of describing and classifying development levels around the world, but some of the most commonly used are **GDP**, **GNP** and the **Human Development Index**.

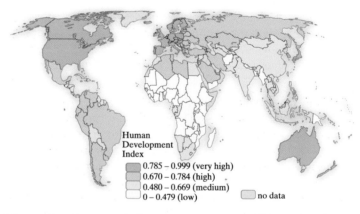

Human Development Index
- 0.785 – 0.999 (very high)
- 0.670 – 0.784 (high)
- 0.480 – 0.669 (medium)
- 0 – 0.479 (low)
- no data

The Human Development Index measures a country's level of development in three basic aspects of human life: health, knowledge, living standards. Health is measured by life expectancy at birth; knowledge is measured by a combination of the adult literacy rate and the combined primary, secondary and tertiary education ratios; living standards are measured by GDP per capita (in US dollars).

The development continuum

The early classification of countries into first/second/third world is now considered too simple. As countries develop, they pass from one condition to another gradually and this process is referred to as the **development**

continuum. Nevertheless, we can still identify particular groupings of countries:

- **Developed countries** are countries with the highest living standards and levels of economic development.
- **Developing countries** are countries at a lower stage of economic development.
- **Least developed countries (LDCs)** have extremely low living standards and associated low life expectancy, poor health and education provision, etc.
- **Newly industrializing countries (NICs)** are those that have achieved development through industrialization in the last 40 years, such as South Korea.
- **Recently industrialized countries (RICs)** are countries such as Mexico which have started this process more recently.
- **Centrally planned economies**, where the total direction and development of a nation's economy is planned and administered by its government, are now rare. North Korea is one example.
- **Oil rich countries**, such as Saudi Arabia and Oman, may have a high GNP per capita, but the wealth is not always distributed evenly.

Globalization

This term describes the process by which the world's economies, societies, and cultures have become integrated through a global network of communication, transportation, and trade.

Good points:

- Supporters of globalization claim that it increases economic prosperity as well as opportunity, especially among developing nations, enhances civil liberties and leads to a more efficient allocation of resources – all countries benefit in the end.

Bad points:

- In many poorer nations globalization is the result of foreign **transnational corporations (TNCs)** taking advantage of the country's lower wage rates to employ a low-cost workforce – commonly in 'sweatshops'. There are even concerns about the emergence of 'electronic sweatshops' due to the outsourcing of service work, such as customer service and IT support, to India. This has undoubtedly resulted in longer hours and an intense pace of work, which is beginning to result in health and social problems.
- On the other hand, opportunities in richer countries have driven talent away from poorer countries, leading to the infamous 'brain drain' which costs the African continent annually over $4.1 billion in the employment of 150 000 expatriate professionals.
- Some people are also concerned about the effect of globalization on culture. Along with the globalization of economies and trade, culture is being imported and exported as well. The concern is that the more dominant nations, such as the USA, may overrun other countries' cultures, leading to indigenous customs and values being lost forever.

Essential notes

Sustainable development: described in the Brundtland Report (1987) as: 'development which meets the needs of the present without compromising the needs of future generations'.

Development gap: refers to the difference in wealth (and associated development indicators such as life expectancy) between developed and developing countries – a gap which seems to be increasing.

North–south divide: the imaginary line separating the 'rich north' from the 'poor south' – often known as the **Brandt line**.

Patterns and processes

The growth of NICs

Newly industrializing countries (NICs) are countries which have experienced very rapid growth in their manufacturing industry since the 1960s. They have benefited from the 'transfer of technology' from incoming TNCs (such as the development of managerial and research skills) and from associated infrastructural developments. This has encouraged the growth of home-based companies like South Korea's Lucky Goldstar, which has itself developed into a fully fledged TNC.

First phase: The most familiar NICs are perhaps the so-called **Asian Tigers**: Singapore, Taiwan, Hong Kong and South Korea. They initially attracted manufacturing industry due to cheap labour costs, the availability of raw materials, weaker environmental and planning laws, cheap land costs and developing domestic markets.

Improved telecommunications technology (satellites and the internet) have made it possible to control production in NIC plants from headquarters in more developed countries, rendering such locations even more attractive.

In the case of South Korea the growth of national companies was encouraged by the government (which owned many companies), but *Chaebol* (large family-owned businesses) also experienced a great deal of growth and have been at the heart of the country's industrial expansion.

Second phase: As these NICs developed, wages rose and Japanese, European and US TNCs began to look elsewhere. South Korea is an example of an NIC that has developed into a country of origin. To remain profitable, labour-intensive industries have moved away to less developed neighbouring countries that have cheaper labour costs than themselves, such as Malaysia, Thailand and Indonesia. The home-grown TNC, Lucky Goldstar, has in fact invested in both developed and developing countries.

South Korea has thus seen rapid and dramatic change:

- It now has the world's 15th largest economy according to GDP; GNP per capita which was $100 in 1963 now exceeds $29 000 (2010). In addition, South Korea rates highly (26th) in the world HDI rankings.
- The economy is heavily dependent on international trade, and in 2009, it was the eighth largest exporter and tenth largest importer in the world. The South Korean government has now started to invest heavily in robotics and aims to become the world's number one robotics nation by 2025.

Third phase: More recently, both China and India have proved attractive targets for **foreign direct investment** (**FDI**) and have experienced rapid and sustained economic growth as a result of this.

China is now the world's second largest economy after the USA in terms of GDP ($4.99 trillion in 2009) and the world's fastest-growing major economy, with an average growth rate of 10% for the past 30 years. China is also the second largest trading nation in the world and the largest exporter and second largest importer of goods – hence, the country's huge impact on the global economy.

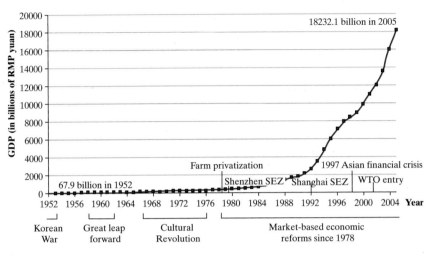

Fig 2
China, nominal GDP, 1952–2005

How has China achieved this?

- 1978 was a pivotal year, marked by a change in government attitude and a decision to move from a centrally planned to a more market-orientated economy under the leadership of Deng Xiaoping. This opened China to TNCs wishing to use the country as an export platform and has made it a major competitor to other Asian NICs.
- Initially, FDI was encouraged in several small '**special economic zones**' (**SEZs**) along the coast. Then in the early 1980s steps were taken to expand the number of areas that could accept FDI with a minimum of red tape accompanied by significant improvements in infrastructure. Fourteen coastal cities (including Shanghai and Shenzen) and three coastal regions were designated 'open areas' for foreign investment.
- A further significant stimulus to growth was provided by China's entry to the **World Trade Organization** (**WTO**) in 2001, allowing better access to global markets providing that China continued to open and liberalize its regimes.

What problems have occurred as a result of China's rapid economic growth?

- Despite the rapid increase in overall GDP, China ranks only 99th in terms of GDP per capita. There is a widening gap in incomes between rich and poor and some of the interior regions such as Sichuan appear to have been 'left behind'.
- Rapid industrialization in the past 30 years has left China with some of the world's worst water and air pollution and widespread environmental damage. Activists have blamed the deadly landslides in Gansu province in August 2010 on years of unchecked development as local governments cut down trees, built roads and developed hydroelectric dams in the name of growth.

Examiners' notes

China is a compulsory case study – you need to be able to explain the reasons for its emergence as an NIC and the positive and negative effects of this growth. A typical question might ask you to describe and comment on the growth illustrated by the graph in **fig 2**.

Essential notes

In the second quarter of 2010, China's economy was valued at $1.33 trillion, ahead of the $1.28 trillion at which Japan's economy was valued. China could become the world's largest economy by 2030.

THE HENLEY COLLEGE LIBRARY

The globalization of services

This can be regarded as the second wave in globalization, following the pattern established by manufacturing industry. Initially, the bulk of services outsourced from one country to another were linked to functions that were repetitive, routine and relatively uncomplicated, such as data processing and customer call centres. However, this itself is changing and today 'off-shoring' has expanded to include more specialized tasks. For example, hospitals are offshoring medical transcription, diagnosis and decisions on surgical intervention, while legal firms are offshoring litigation and patent research. In recent years one of the most attractive locations for the outsourcing of services has proved to be India and, in particular, the state of Karnataka which houses Bangalore, often known as India's 'silicon city'.

Why India and why Karnataka?

- Low cost, but high quality and adaptable labour force
- Second largest English-speaking human resource in the world
- World's third largest 'brain bank' with around 2.5 million technical professionals
- Investment-friendly and supportive government policies within a democratically stable country
- Well-developed power, transport and communications infrastructure – in line with other developed countries
- Access to a large and growing local market with a burgeoning 'middle class' of some 250–350 million people with increasing purchasing power
- A well-developed banking system
- Legal protection for intellectual property rights (IPR)

Ten per cent of Indian graduates are produced in Karnataka, which has, historically, been a place for technology and R&D-based institutions. This was the first state to set up engineering colleges and a university of technology. It has the best telecoms infrastructure in the country. Software technology parks exist at Mysore, Hubli and Mangalore with high speed data communications facilities.

A specialized industrial park, Electronics City, has been built at Bangalore – one of India's largest, spread over 1.3 km². Bangalore is centrally located, and enjoys a favourable climate, excellent social, health and education facilities and a cosmopolitan reputation.

However, the country's standing as a premier offshoring destination with a booming economy often masks the fact that 70% of its population lives in rural areas. Despite its reputation for ICT, only 7% of India's population had access to the internet in 2008 and 400 million Indians have no access to electricity whatsoever.

Growth in the 21st century

Many **emerging markets** have achieved growth by moving from centralized political and economic control to more open market systems, relaxing controls on foreign exchange, removing trade barriers and allowing greater autonomy for financial institutions. This has not only stimulated growth by attracting FDI but has generated rising living standards and increased opportunities for local populations.

China and India continue to rank highly in the 2010 FDI Confidence Index (a regular survey conducted by management consultancy firm AT Kearney). China has led the index rankings since 2002 whereas India has dropped back to third place since 2007, displaced perhaps surprisingly by the USA. This may reflect investors' 'flight to safety' in a time of economic turmoil.

In all, 12 of the most attractive 25 FDI destinations are developing or 'emerging' markets. After China, India and the USA, Brazil is 4 in the list; Mexico 8; the United Arab Emirates 11; Vietnam 12; and other Gulf States (Bahrain, Kuwait, Oman and Qatar) 15.

Russia fell sharply in the rankings in 2010 (from 9 to 18) with a particularly severe economic slump, worse than that experienced by any of the other **BRIC countries**. Despite the huge opportunities in the oil and gas sector, it seems that transparency remains a major obstacle for foreign investors.

On the other hand, Saudi Arabia's efforts to improve its FDI climate with a series of economic reforms appear to be paying off as it joined the top 25 for the first time. The Saudi government's attempt to diversify away from the petrochemical industry is attracting foreign investors, particularly in telecommunications and utilities.

Egypt also entered the top 25 for the first time in 2010. The country has long been an FDI destination because of its oil and gas resources; a privatization scheme; and, increasingly, its large consumer market. Egypt, like Saudi Arabia, has implemented a number of reforms to improve its business environment.

Essential notes

The term 'emerging markets' refers to nations experiencing rapid growth and industrialization. Currently, there are 28 emerging markets in the world, with China and India leading the way. The term is considered by some to be somewhat dated, but a new term has yet to gain any credence.

Essential notes

The BRIC countries – Brazil, Russia, India, and China – are all deemed to be at a similar stage of newly advanced economic development. The acronym was coined by Jim O'Neill in a 2001 paper entitled 'The World Needs Better Economic BRICs'. .

Examiners' notes

An excellent slide show detailing the top 25 countries for FDI in 2010 is available at: http://images.businessweek.com

This could help you to build up a case study of one of these countries.

Fig 3
The world's emerging markets

Aspects of globalization

TNCs: characteristics and spatial organization

A transnational corporation (TNC) is a firm which has the power to coordinate and control operations in more than one country. Such organizations are hierarchical and usually have a recognizable home base incorporating the headquarters and research and development (R&D) arm in developed countries, with centres of production overseas.

One example is WalMart, the world's number one TNC, employing 2.1 million people worldwide with 7 500 stores, generating sales of $408.21bn in 2009. TNCs such as WalMart have a high level of diversification and are multi-product companies. Other TNCs produce **global products** (sold all over the world in the same form), e.g. Coca Cola.

Impacts of TNCs	
Positives	**Negatives**
• TNCs are a vital source of FDI – in the UK in 2007, FDI generated over 50 000 jobs. • TNCs stimulate the multiplier effect. The company itself may require locally produced components and other supply and distribution services. Meanwhile, increased wealth and disposable incomes will generate domestic demand and stimulate further growth. • TNCs not only provide employment, they increase local skills. In some cases, this may help to offset large-scale unemployment caused by the mechanization of agriculture. • They are often responsible for the transfer of technology such as '**just in time**' **production**. • They may construct or improve local infrastructure such as roads and bridges, which benefits not only the company, but the local area overall.	• TNCs can prove lethal competition for local firms which may go out of business, creating local hostility. • TNCs often face negative attitudes from local authorities, residents and environmentalists. The Chinese car industry has received heavy investment from Ford and VW, but this has been partly responsible for increasing pollution levels and contributing to growing traffic problems in cities such as Shanghai. • Many of the jobs offered are low skilled. Managerial positions are often filled by people who have moved with the TNC, providing little prospects for locals to develop within their jobs and gain promotion. • Some TNCs stand accused of exploiting cheap, flexible, non-unionized labour in sweatshops in developing countries. • Like the tourist industry, TNCs can be fickle employers – moving elsewhere in the interest of profitability with little concern for locals. Much of the capital generated will, in any case, find its way back to the country of origin.

Reasons for the growth of TNCs

- TNCs have control over terms of trade and may even be able to reduce the effects of quota restrictions on the movement of goods by locating in their intended market.
- TNCs can take advantage of government policy on a global scale. They can exploit differences in the availability of capital, tax levels, subsidies and grants, environmental controls, cheap labour, land and building costs. For example, Dyson moved production from Wiltshire to Malaysia in 2002 to take advantage of much lower labour rates.

Case study: Nike

Nike employs more than 33 000 people globally, but the company headquarters are located in Beaverton, Oregon, USA. Nike sponsors many high-profile athletes and sports teams around the world, with the highly recognized trademarks of 'Just do it' and the Swoosh logo. In 2009, Nike reported record revenues of $19.2bn, a 3% increase over the previous year's earnings.

Nike has been much criticized for contracting with 'sweatshop' factories in NICs such as Vietnam, Indonesia and Malaysia. A July 2008 investigation by the Australian broadcaster Channel 7 News found a large number of cases involving forced labour in one of the largest Nike clothing factories in Malaysia, with workers living in squalid conditions. Nike subsequently stated that they would take corrective action to ensure that continued abuse did not occur. In fact, in 2010 Nike was recognized as one of the world's most ethical companies by The Ethisphere Institute. The Institute recognizes organizations that promote ethical business standards and practices by going beyond legal minimums, introducing innovative ideas benefiting the public, and forcing their competitors to follow suit – suggesting that Nike has done much to improve its image.

The textile industry is often accused of having a negative environmental impact with its huge demands on water resources and its fossil fuel and raw material consumption. However, Nike tries hard to counteract these assertions with a number of different projects. One campaign that Nike began for Earth Day 2008 was a commercial that featured Steve Nash wearing Nike's Trash Talk shoe, a shoe that had been constructed in February 2008 from pieces of leather and synthetic leather waste that derived from the factory floor. The Trash Talk shoe also featured a sole composed of ground-up rubber from a shoe recycling programme. However, only 5 000 pairs were produced for sale.

Nike was named in the top 10 of Newsweek's 2009 Green Rankings (which ranks 500 of America's largest corporations) citing Nike as an industry leader in environmental management of suppliers. Another project the company has begun is called Nike's Reuse-A-Shoe programme. Started in 1993, this is Nike's longest-running programme benefiting both the environment and the community by collecting old athletic shoes of any type in order to process and recycle them. The material that is created from the recycled shoes is then used to help create sports surfaces, such as basketball courts, running tracks, and playgrounds.

Essential notes

Find out about the Nike Grind and Reuse-a-Shoe programme at: www.nikereuseashoe.com

The Nike Foundation works to improve the lives of adolescent girls in developing countries: www.nikefoundation.org

Essential notes

Bilateral aid is aid given directly from one country to another, in the form of money, goods or services.

Multilateral aid is aid which comes from several different countries – often through international agencies such as the World Bank.

NGOs are non-governmental organizations, including charities such as Oxfam, which provide both money and professional support. This type of aid is less likely to come with any conditions attached and is perhaps the most successful in terms of reaching the people who need it most.

Development issues

Trade and aid

There is an enduring debate about aid and trade. It could be argued that trade is better because this helps the economy of a country by increasing the volume and value of exports. This in turn will help create jobs, which will generate income. Eventually, some of this income will be saved and will generate domestic demand and investment which will lead to more growth – the multiplier effect. In addition to raising living standards, local people will gain skills and expertise.

However, this model assumes that developing countries will all go through the same cycle that the developed countries went through in the 19th and 20th centuries. In support of this argument is the experience of the NICs, which do indeed appear to have followed a similar path to economic development. On the other hand, these countries had largely stable governments, a relatively well-educated workforce, and they employed strong protectionist policies to stimulate growth.

Aid may well be necessary to assist the world's least developed countries, which have to face multiple challenges in terms of political instability, war, drought and the battle against HIV/AIDS. Aid may come in the form of capital or it may come in the form of goods or technical assistance to create infrastructure or build up agriculture and industry. However, it can also create high levels of inflation (if the aid is used to fund the current account budget deficit) and it can lead to aid dependency. Aid does not create sustainable growth or development and food aid in particular may depress the prices in the domestic market.

Furthermore, aid does not always reach the people for whom it was originally intended – corruption is a problem in many receiving countries. The directors of several relief organizations in Myanmar claimed that some of the international aid arriving into the country for the victims of Cyclone Nargis in 2008 was stolen, diverted or warehoused by the country's army.

Tied aid (aid with certain conditions) will limit the sovereignty of nations and may eventually cause resentment.

Short-term aid deals with emergencies such as the tsunami that devastated parts of Asia on 26 December 2004. This type of aid brings immediate help to people – flying in food to prevent starvation, tending the injured and sick, and trying to prevent the spread of disease.

Long-term aid is required where problems are deep-rooted. For example, the climate in some parts of sub-Saharan Africa means that drought is a common occurrence. This requires long-term development projects to try to prevent water shortages – for example, sinking wells and providing villages with water pumps to enable permanent access to underground water supplies.

Top-down aid is where large-scale projects such as dam building are directed 'from the top', often by organizations such as the World Bank.

Bottom-up schemes, also known as grassroots initiatives, are often funded by NGOs working closely with local communities, using local ideas and skills.

Case study: WaterAid in Burkina Faso

Burkina Faso is a small, landlocked country in west Africa with a population of 15.3 million. It is one of the poorest in the world, having one of the world's lowest GDP per capita figures: $1 200. Agriculture represents 32% of its GDP and occupies 80% of the working population. It is ranked 174 out of 177 countries by the UN and one in five children die before their fifth birthday. There are few schools, health facilities and public services, especially in rural areas. Just over half of the population have access to clean water, while less than 15% have access to sanitation.

Average life expectancy at birth in 2004 was estimated at 48 for females and 47 for males, and it is estimated that there were as few as six physicians per 100 000 people at this time. In addition, there were only 41 nurses, and 13 midwives per 100 000 people. According to the WHO in 2005, an estimated 72.5% of Burkina Faso's girls and women had suffered female genital mutilation.

The country (originally known as Upper Volta) regained its independence from France in 1960 and was renamed Burkina Faso ('the land of the upright people') in 1984. However, it has spent many of its post-independence years under military rule and has suffered several military coups. Unemployment causes a high rate of emigration. For example, 3 million citizens of Burkina Faso live in neighbouring Ivory Coast (Côte d'Ivoire).

Reducing rainfall levels and recurring droughts are a serious issue. The Sahel region in the north of the country typically receives less than 600 mm of rainfall a year with high temperatures which can soar to 47 °C. As a result, large numbers of the rural poor, who traditionally rely on farming for a living, are moving to towns and cities looking for work. Hence the number of unplanned squatter settlements, without services like water and sanitation, are growing.

WaterAid and its partners carry out a range of projects. For example, in Ramongo, hygiene education is under way along with the construction and rehabilitation of boreholes. In the Koulpeologo province WaterAid is developing the WASH concept (water, sanitation and hygiene for all) in schools. This involves training teachers and encouraging the creation of school health clubs. These are examples of grassroots initiatives.

Soap production is another innovative scheme, where women make and sell soap. This is enabling women to make money while increasing hand washing, a simple practice that can reduce diarrhoeal diseases by over 40%. WaterAid's partners invest funds for the initial set-up costs, which the women refund once the soap is produced and sold.

BURKINA FASO

Fig 4
Burkina Faso, West Africa

Examiners' notes

This case study illustrates many of the challenges faced by the world's LDCs, particularly in terms of poor quality of life and social problems. Burkina Faso was also one of the first countries to qualify for debt relief under the HIPC initiative, totalling $700m.

Essential notes

According to WaterAid, at the current rate of progress, the UN will not meet its goal to bring sanitation to the places that need it most, primarily sub-Saharan Africa, until the 23rd century – 200 years late. Find out about their latest campaign: 'Dig Toilets, not Graves'.

Essential notes

In 2009, net bilateral **ODA**
(official development
assistance) to Africa was
$27bn, representing an
increase of 3% in real terms
over 2008. $24bn of this aid
went to sub-Saharan Africa,
an increase of 5.1% over 2008.
(OECD figures)

Countries at very low levels of economic development

Since 1968, the term 'least developed country' (LDC) has been applied to a
country which, according to the **United Nations** (**UN**), exhibits the lowest
HDI ratings of all countries in the world. A country is classified as an LDC if
it meets three criteria:

- Low incomes – three-year average **GNI** per capita of less than
  US$905, which must exceed $1 086 for countries to leave the list
- Human resource weaknesses based on indicators of nutrition,
  health, education and adult literacy
- Economic vulnerability, indicated by factors such as: low levels
  of economic diversification and annual per capita energy
  consumption; instability of agricultural production; instability
  of exports of goods and services; and even the percentage of
  population displaced by natural disasters

In addition, many LDCs also face challenges posed by civil war and ethnic
conflicts, political corruption and instability.

Quality of life

One of the most frequently used measures of poverty is the percentage of
people who live on less than one US dollar a day (increased by the UN in
2009 to $1.25). About one fifth of the world's population are in this position
– in 2005, nine of the ten countries with the largest percentage of people in
this category were in Africa: Madagascar, Sierra Leone, Burundi, Gambia,
Niger, Zambia, Central African Republic, Nigeria and Mali had between
49% and 73% of their populations living in extreme poverty. Many of these
countries seem to be stuck in a vicious circle of poverty (as illustrated
below) and are still very dependent on external finance (aid).

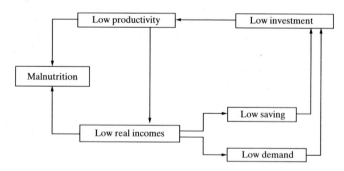

Fig 5
The vicious cycle of poverty

Debt

LDCs appear to be facing an ever increasing debt burden. This began with
rising oil prices in the 1970s and was exacerbated by higher interest rates
in the 1980s coupled with falling export prices (particularly of primary
products) and, in many cases, poor economic management. Some would
argue that the problem is even more deeply rooted and that many of the
world's poorest countries have started their independent status with heavy
debt burdens imposed by the former colonial occupiers. As such, it could
be argued that the world's richer countries now have a moral obligation to
help out.

In response to this kind of thinking, and with the view that there was in any case little hope of any of these debts being repaid, the **International Monetary Fund** (**IMF**) and the World Bank produced the **Heavily Indebted Poor Countries** (**HIPC**) initiative in 1996. The aim was to ensure that no poor country faced a debt burden it was unable to manage.

UN Millennium Development Goals

For the world's developed nations, poverty, child mortality, poor maternal health, unsafe drinking water and rampant disease are, fortunately, shadows from the past. However, these are serious issues that continue to plague LDCs. With these factors in mind, world leaders adopted the UN Millennium Declaration in September 2000. There are eight specific goals to meet before the 2015 deadline.

The UN Millennium Development Goals (MDGs) are:

1. The eradication of poverty and hunger
2. Universal access to primary education
3. Promotion of gender equality and empowerment of women
4. Reduction in child mortality
5. Improvements in maternal health
6. Combating disease, particularly HIV/AIDS and malaria
7. Environmental sustainability
8. Developing a global partnership for development

In 2005, to help accelerate progress toward the United Nations **Millennium Development Goals** (**MDGs**), the HIPC initiative was supplemented by the **Multilateral Debt Relief Initiative** (**MDRI**). However, in order to qualify, certain criteria have to be met. Countries must:

- Be eligible to borrow from the World Bank's International Development Agency and the IMF's Extended Credit Facility – these provide interest-free loans and grants to the world's poorest countries
- Face an unsustainable debt burden that cannot be addressed through traditional debt relief mechanisms
- Have established a track record of reform, and sound policies through IMF- and World Bank-supported programmes
- Have developed a **Poverty Reduction Strategy Paper** (**PRSP**) and develop and implement these reforms for at least a year.

By 2010, of the 40 countries eligible or potentially eligible for HIPC initiative assistance, 30 were receiving full debt relief from the IMF and other creditors. Six other countries were close to meeting the criteria and were receiving interim debt relief, while four countries had yet to do so: Eritrea, Sudan, Somalia and the Kyrgyz Republic.

Essential notes

Before the HIPC initiative, these countries were spending slightly more on debt servicing than on health and education combined. However, spending on such vital services is now, on average, about five times the amount of debt–service payments.

Essential notes

In the foreword to the MDG Report 2010, UN Secretary General Ban Ki-moon states that: 'It is clear that improvements in the lives of the poor have been unacceptably slow, and some hard-won gains are being eroded by the climate, food and economic crises'. However, the report also cites big gains in cutting the rate of extreme poverty, getting children into primary schools, addressing AIDS, malaria and child health, and the real prospect of reaching the MDG target for access to clean drinking water.

Essential notes

For information on the Rio Earth Summit's sustainability principles visit: www.un.org/geninfo/bp/enviro.html

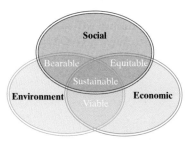

Fig 6
The 'three pillars' of social, environmental and economic sustainability

Essential notes

Visit the Sustainable Tourism Gateway for more information on this topic: www.gdrc.org/uem/eco-tour/eco-tour.html

and the website of the International Ecotourism Society: www.ecotourism.org

For case studies on ecotourism, visit: http://geography.about.com/od/locateplacesworldwide/a/ecotourism.htm

Economic vs environmental sustainability

The UN's widely quoted definition of sustainability dates from the 1987 Brundtland Commission and is as follows: 'Sustainable development is development that meets the needs of the present without compromising the ability of future generations to meet their own needs'. At the 2005 World Summit it was noted that true sustainability requires the reconciliation of environmental, social and economic demands: the 'three pillars' of sustainability (see **fig 6**).

The earth has a finite capacity to provide resources and to absorb waste; many would say that human demands already exceed this capacity. Current lifestyles in the developed world, to which many people in the developing world quite reasonably aspire, are unsustainable. The UN has stated, in its Millennium Declaration, that 'current unsustainable patterns of production and consumption must be changed'.

For many environmentalists the idea of sustainable development is an oxymoron, as development seems to necessitate environmental degradation. The challenge for sustainability is to curb and manage Western consumption while raising the standard of living of the developing world without increasing its resource use and environmental impact.

There is also the social challenge, which entails examining the relationship between human rights and human development, corporate power and environmental justice, global poverty and global citizenship. War, crime and corruption divert resources from areas of greatest human need, damage the capacity of societies to plan for the future, and generally threaten both human well-being and the environment.

Sustainable tourism: myth or reality?

Sustainable tourism is an industry which attempts to have a low impact on environments and local cultures, while helping to generate income and employment, and contribute to the conservation of local ecosystems. It is responsible tourism that is ecologically and culturally sensitive – also known as **ecotourism**. However, critics say that increased tourism to sensitive areas or ecosystems without proper planning and management can actually harm the ecosystem and its species because the infrastructure needed to sustain tourism such as roads can contribute to environmental degradation. Ecotourism is also said to have a negative impact on local communities because the arrival of foreign visitors and wealth can cause resentment. It can make the area dependent on tourism which can be a fickle employer, especially if there is political unrest.

Global social and economic groupings

The key terms are:

- **Free trade areas** – countries within such areas trade freely without the imposition of quotas or tariffs, but maintain restrictions on goods coming from outside the area. An example would be NAFTA – the North American Free Trade Association.
- **Customs unions** – members impose a tariff on imports from outside the group – e.g. Mercosur, the South American Regional Trade Agreement.

- **Common markets** – similar to customs unions, but also allowing the free movement of labour and capital. The EU used to exist in this form.
- **Economic unions** – these incorporate all of the above, but also agree to adopt common policies on issues such as agriculture, energy, transport, pollution, etc. The European Union is an example.

The European Union

The **EU** is an economic and political union of 27 member states with over 500 million citizens, committed to regional integration. It was established by the Treaty of Maastricht in 1993 upon the foundations of the European Economic Community. The EU has developed a single market through a standardized system of laws which apply in all member states, and ensures the free movement of people, goods, services, and capital, including the abolition of passport controls by the Schengen Agreement between 22 EU states. It enacts legislation in justice and home affairs, and maintains common policies on trade, agriculture, fisheries and regional development. Seventeen member states have adopted a common currency, the euro, constituting the **eurozone**.

Consequences of international groupings	
Positives	**Negatives**
• Increased trade opportunities with reduced trade barriers • Ease of movement for workers between member countries • Ease of movement for travellers and workers with the removal of passport/customs checks • Support for struggling sectors of the economy, e.g. via the EU's Common Agricultural Policy • Support for declining industrial regions, including urban regeneration • Protection against cheap imports • Wider transport and environmental policies • Possibility of developing a common currency, preventing currency fluctuations and simplifying transactions • Reduction in the risk of war between member countries	• An increased feeling of centralized government and an associated loss of sovereignty • Ultimately, such groupings may actually foster separatism as smaller enclaves such as the Basques feel even more disenfranchised • Loss of financial control to a central body such as the European Bank • Increased paperwork • Pressure to adopt certain legislation such as the European Social Chapter/Working Time Directive • Difficulty in meeting requirements and costs of EU directives (e.g. on pollution emissions) • Conflict can result from agreements to share resources, as between UK and Spanish and French fishermen

Examiners' notes

The specification requires a focus on the EU here, but it would be a good idea to investigate two or three of the other major trade blocs in the world to show breadth of knowledge and understanding.

Choose from the following:

- NAFTA
- ASEAN
- SADC
- MERCOSUR
- CARICOM

Essential notes

Here is a timeline for the EU:

1957: Treaty of Rome created the European Economic Community (EEC), initially six countries: France, (West) Germany, Italy, Belgium, Luxembourg and the Netherlands

1973: UK, Ireland and Denmark joined

1981: Greece

1986: Spain and Portugal

1995: Austria, Finland, Sweden

2004: Cyprus, Czech Republic, Estonia, Hungary, Latvia, Lithuania, Malta, Poland, Slovakia, Slovenia

2007: Bulgaria, Romania

Causes and expression of conflict

In political terms, '**conflict**' refers to an ongoing state of hostility between two or more groups of people. In a geographical context, conflict is often the result of opposing views about the ways in which a resource might be developed or used. In some cases, conflict may be resolved peacefully, whereas in others the ill-feeling may escalate into full-blown war.

The Uppsala Conflict Data Program (UCDP) is a university-based data collection programme which continuously gathers information on armed conflict and other types of organized violence. It also includes descriptive information on the causes of such conflicts and of one-sided violence, along with brief descriptions of rebel groups, governments and related items. The table below shows some of the conflicts. This already gives a taste of the wide variety of conflicts that exist around the world, in terms of both their nature and their origin.

Essential notes

Uppsala Conflict Data Program (UCDP): www.pcr.uu.se/research/UCDP/.

Start of conflict	War/conflict	Location	Cumulative fatalities
2003	Civil war in Iraq	Iraq	100 000–1 000 000
2004	War in northwest Pakistan	Pakistan	30 452
2006	Mexican drug war	Mexico	More than 28 000
2009	Sudanese nomadic conflicts	Sudan	2 500

Table 1
Some of the global conflicts of the 21st century

Causes of conflict

Identity: This refers to a sense of belonging to a particular group or geographical area, characterized by its ethnicity, language and/or religion. Within this category are several sub-categories:

- **Nationalism** – loyalty and devotion to a nation.
- **Regionalism** – loyalty to a distinct region. Some believe that strengthening a region's governing bodies and political powers within a larger country helps to develop a more rational allocation of the region's resources for the benefit of the local populations.
- **Localism** – an affection for a particular place. This again may generate conflict, usually on a smaller scale, such as the Lune Valley residents who are objecting to proposals to build one of the northwest's tallest wind farms two miles outside Kirkby Lonsdale. Such objections are sometimes described as 'nimbyism' (derived from 'not in my back yard').
- **Ethnicity** – this is the grouping of people according to their ethnic origins or characteristics.
- **Culture** – refers to the customary beliefs, social norms and traits of a racial, religious or social group. Such cultural variations can certainly enrich an area, but they can also give rise to conflict. The Basque conflict (pp. 86–87) is based on both ethnic and cultural differences.

Territory: Territorial disputes concern geographical areas belonging to, or under the jurisdiction of, governmental authorities. Conflict can occur when there is dispute about who has authority over a particular area, as in

Examiners' notes

Note that this section of the specification does not require a case study, but it would be a good idea to be able to give a brief example under each of these headings.

the case of Western Sahara, which has been on the United Nations list of non-self-governing territories since 1963. Morocco and the Sahrawi Arab Democratic Republic (SADR) dispute control of the territory.

Ideology: An ideology is a set of ideas that constitute an individual's or a group's goals, expectations and related actions. For example, there is a huge difference between the Western democratic style of government and the single-party state of North Korea, which conforms to the Juche ideology of self-reliance, developed by the country's former president, Kim Il-Sung.

Expression of conflict

Non-violent: This type of conflict is generally expressed through debate or perhaps some form of protest such as a march or a petition. Non-violent protests can be very successful, such as the Orange Revolution in Ukraine between November 2004 and January 2005. This occurred as a result of a fraudulent presidential election marred by corruption and voter intimidation. Kiev, the Ukrainian capital, was the focal point of the movement with thousands of protesters demonstrating daily. The protests succeeded when the results of the original election were annulled, and a new vote was ordered culminating in a clear victory for Victor Yushchenko, who was duly elected president. On the other hand, the Tiananmen Square protests of 1989, led mainly by students and intellectuals, were ultimately quelled by the Chinese military, leaving more than 2 500 dead and attracting intense international criticism.

Political activity: The UK has a wide range of political parties, including national parties in Scotland, Wales and Northern Ireland. These parties have their own policies and ideologies, and compete for the support of the public to win power. In parliament, the party with the most Members of Parliament forms the government. The opposition parties contribute to policy and legislation through constructive criticism and debate, oppose government proposals that they disagree with, and put forward their own policies.

Terrorism: 'Terrorism' is a highly politicized term, which is defined and viewed in very different ways by different groups. It has proved an effective tactic for the weaker side in some conflicts, offering a degree of coercive power at a fraction of the cost of military force. Due to the secretive nature and small size of terrorist organizations, they often offer opponents no clear organization to defend against or to deter. A particular present-day challenge for state authorities is the tactic of suicide bombing.

Insurrection: Those engaged in insurrection are called insurgents, and typically use guerrilla tactics against the armed forces of an established regime. The Provisional Irish Republican Army (IRA) waged a guerrilla campaign against British rule in Northern Ireland between 1969 and 2005. Over this time period, the IRA is believed to have been responsible for the deaths of approximately 1 800 people.

War: War is a phenomenon of organized violent conflict between nations or sometimes within nations (civil war) and is characterized by extreme aggression, serious disruption to societies and economies and high rates of mortality (see **table 1**).

Conflict over the use of local resources

In the UK, conflicts over local resources (e.g. land, buildings, space) are resolved by market processes or planning processes or sometimes a combination of the two.

Market processes operate in an environment where the ability to pay the going rate takes precedence over local and national concerns. It is often the case that objectors cannot afford to outbid the developers and the development may go ahead with the minimum of consultation. Market processes are quick, as they concern only the owner, the purchaser and anyone else involved in the transaction – lawyers, etc. Although there may be the opportunity to voice objections, there is no right to appeal.

Planning processes attempt to provide a means by which planners can listen to the local community, listen to the organization responsible for a proposed change, and ultimately have control of the development concerned. Planning enquires are expensive and time consuming. Different viewpoints are heard, but they may be biased towards those who can afford to present their case most strongly. Any refusal to grant planning permission by a local authority committee may lead to an appeal, or may result in the developer going to a higher body for arbitration, such as the Department for the Environment, Food and Rural Affairs (Defra). Plans may then be approved subject to various modifications which may placate the opposition.

The typical planning process operates as in the diagram below.

Examiners' notes

Remember that planning committees such as that of Wakefield Council will need to weigh up the pros and cons of any proposals and consider the wider benefits of any scheme as opposed to any local objections.

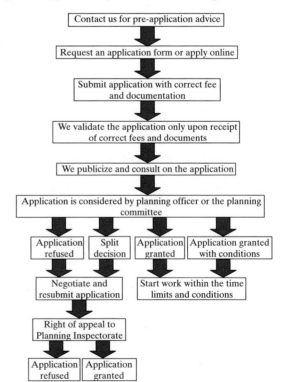

Fig 1
Wakefield Council's planning process

Case study: Knock Rushen, Isle of Man

Conflict here arose as a result of a proposal to build 41 homes and four apartments on the urban/rural fringe of Castletown and adjacent to Scarlett, a local beauty spot and wildlife reserve. The protest group Save Our Scarlett (SOS) successfully campaigned against development on the land on numerous occasions over a period of some 13 years. Ultimately, however, the builder's perseverance paid off and local developers Hartford Homes were granted approval for the scheme in 2006.

Parties in favour of the development included some of the local shopkeepers, estate agents, landowners and potential purchasers. An important consideration was that the growth of Castletown is restricted in other directions due to its proximity to the coast, safety regulations related to Ronaldsway airport to the north, and the reluctance of landowners to sell.

On the other hand, the Manx Nature Conservation Trust and Manx National Heritage were opposed to development in the area. Scarlett is considered a unique local beauty spot, home to birds such as the chough and grey plover, while Knock Rushen itself is thought to conceal an ancient Viking burial site. Many residents also voiced their concern that the development would prove 'the thin end of the wedge' and would pave the way for even more building. This could put pressure on local services such as the Victoria Road primary school.

There are only two main routes out of Castletown – firstly via the town square, where the road is very narrow and cannot be widened due to preservation orders on old buildings such as Castle Rushen itself dating from 1090; and secondly, through a residential area which is already congested due to the number of cars parked on narrow roads, such as The Crofts. A third option, via the bypass, would entail passing two local schools, Castle Rushen High School and The Buchan, an area already highly congested and dangerous due to school traffic at peak times.

One of the other arguments against the proposal was the need to utilize brownfield sites within Castletown before developing on greenfield sites, and this was an argument which held sway for some time. However, in the interim, the main brownfield site available at Farrants Way was developed for apartments by another local company, Dandara. Meanwhile, Hartford Homes agreed that the Viking burial mound should be protected and fenced off as a 'feature', in an effort to meet some of the local concerns. As a result of this amendment and the agreement to keep the development relatively small scale and in keeping with the local environment, Hartford Homes won planning approval in 2006 and the development was almost complete in late 2010.

Examiners' notes

This section of the course does require a case study. You must understand the reasons for the conflict and the attitudes of the various groups involved. Think about the impact when the conflict is resolved – who wins and who loses? What processes were used to resolve the conflict?

Essential notes

Visit the Hartford Homes website at www.hartford.co.im to view this development for yourself. Alternatively, there are many other local conflicts worth investigating, such as the objections to a third runway at Heathrow or the proposed wind farm at Kirkby Lonsdale.

The geographical impact of international conflicts
Case study: East Timor (Timor-Leste)

Fig 2
East Timor and its neighbours

Background: East Timor, or Timor-Leste as it is now named, forms the eastern section of the island of Timor, which lies 1 000 miles south of the Philippines and 400 miles northwest of Australia. The western part of the island is under Indonesian rule; East Timor was a Portuguese colony for some 400 years before declaring independence in 1975. However, later that year it was invaded and occupied by Indonesia; in 1976 it was declared Indonesia's 27th province. Indonesia relinquished control of the territory in 1999 following a UN-sponsored referendum, and East Timor, or Timor-Leste, became the first new sovereign state of the 21st century on 20 May 2002.

During the years of Indonesia's occupation, 250 000 Timorese and 10 000 Indonesians were killed. One particularly infamous event was the Santa Cruz massacre, which took place on 12 November 1991. More than 270 Timorese mourners were killed by Indonesian troops as they attended a memorial service marking the death of a young man also killed by the Indonesian military. The East Timorese guerrilla force, Falintil, fought a vigorous campaign against the Indonesian forces during the occupation, some members having been trained in Portugal by Portuguese special forces.

Economic impacts of the conflict: In late 1999, about 70% of the economic infrastructure of East Timor was destroyed by Indonesian troops and anti-independence militias, carrying out a 'scorched-earth' policy. Some 260 000 people fled westwards. The immediate economic consequence of the crisis was felt in terms of the disruption of the flow of people and goods into and out of the capital, Dili, and between East and West Timor. This led to shortages of, and higher prices for, many essential commodities including food and fuel. Over the longer term, the dislocation of people caused by the crisis has had a negative impact on agricultural production and incomes. In addition, it has affected the government's tax and revenue base. For example, there has been a marked decrease in fee collection for services such as electricity and water.

On the positive side, the crisis has prompted the government to increase and accelerate budget expenditure for employment generation projects and housing reconstruction. There has also been significant short-term economic gain from the influx of international support and the establishment of the new UN Integrated Mission in Timor-Leste (UNMIT). For example, between 2002 and 2005, an international programme led by the UN substantially reconstructed the infrastructure and, by mid-2002, all but about 50 000 of the refugees had returned.

Overall, however, rebuilding business confidence is likely to be difficult until a durable solution to the crisis is found. Continued outbreaks of violence, along with the exposure of extreme underlying state instability, will likely curtail the prospect of significant new foreign investment for some time to come.

Social impacts of the conflict: There are also deep-rooted social problems in East Timor, many of which are consequences of the economic problems. The 15 major gangs claim a membership of 90 000, almost one tenth of the population. The two largest gangs are 7/7 and PSHT. Gang culture means that although Indonesia has withdrawn, there is still a large amount of violence and the nation is still in a state of disrepair. Some 2 000 homes in Dili were destroyed by the Indonesian military, and the deliberate destruction of infrastructure left 50% of the Timorese population without access to clean drinking water and 60% lacking adequate sanitation.

At the height of the conflict, the infant mortality rate peaked at about 170/1 000 though this has now fallen to 75/1 000; malnourishment, malaria and dengue fever are still issues. The conflict left the country with a crumbling education system as Indonesian secondary school teachers left with the Indonesian military.

Environmental impacts of the conflict: These revolve around the flight of 80 000 people to rural areas. This put severe pressure on subsistence farmers and the already exhausted land. Food security became an increasing problem. One third of the agricultural land in East Timor is degraded due to a combination of poor soil, deforestation, overgrazing and slash-and-burn agriculture. Another environmental casualty of the occupation was East Timor's forests. In 1975, about 50% of East Timor's land was primary and secondary forest. By 1999, only 1% of these forests remained. Most of the deforestation was conducted under logging operations for teak, redwood, sandalwood and mahogany for export. The use of wood as a primary fuel source has added to the problem of diminishing forests and has had repercussions in terms of soil erosion and localized flooding.

Essential notes

UNMIT is a multidimensional, integrated UN peacekeeping operation, which was established in the wake of the major political, humanitarian and security crisis which erupted in Timor-Leste in 2006. UNMIT has been mandated to support the Timorese government in 'consolidating stability, enhancing a culture of democratic governance, and facilitating political dialogue among Timorese stakeholders, in their efforts to bring about a process of national reconciliation and to foster social cohesion'. See: www.un.org/en/peacekeeping/missions/unmit/

Examiners' notes

Note that this section has deliberately been divided into economic, social and environmental impacts. It is essential that whichever case study you use, you should be able to provide this sort of detail for at least one international conflict. You would be well advised to study more than one, though perhaps not in such great detail.

Essential notes

Multicultural society:
A social grouping which contains members from a wide variety of national, linguistic, religious or cultural backgrounds.

International migration: The movement of people across national frontiers involving a permanent change of residence (for at least one year, according to the UN definition).

Refugee: Defined by the UN in 1952 as 'someone who, owing to a fear of being persecuted for reason of race, religion, membership of a particular social or political group, is outside the country of his/her nationality and is unable or, owing to such fear, is unwilling to return to that country.' This definition has now been widened to include those fleeing from civil war, ethnic, tribal or religious violence, and environmental disasters such as earthquakes, volcanoes and famines.

Ethnic segregation: The clustering together of people with similar ethnic or cultural characteristics into separate urban residential areas.

The challenge of multicultural societies in the UK

Multicultural societies have formed as the result of migration. Significant migrations affecting the UK are as follows:

19th century	Early 20th century	Late 20th century	21st century
Jews escaping persecution in Russia and Poland. Jews were targeted under Tsarist rule and banned from many jobs and locations. They also suffered anti-Jewish riots, known as pogroms.

Irish escaping poverty in Ireland (including the Irish potato famine, 1846–1850). | 1930s/40s: Jews and Poles escaping fascism and the Second World War (Holocaust, 1941 to 1945).

1948–60s: Caribbean workers were invited to help rebuild post-war Britain, mainly working in public services. Employers such as British Rail, the NHS and London Transport recruited almost exclusively from Jamaica and Barbados.

It is estimated that about a quarter of a million people of Afro-Caribbean origin settled permanently in Britain between 1955 and 1962.

1950s–60s: Many immigrants also arrived from India, Pakistan and Bangladesh hoping to escape poverty, and seeking work in public services and the large textile industries of Yorkshire and Lancashire. | 1970s: Ugandan East African Asians sought to escape persecution under President Idi Amin's regime or were forcibly expelled in 1972.

1975 onwards saw a wave of Vietnamese immigration after the end of the Vietnam War. Most early immigrants were refugee 'boat people' fleeing persecution by the victorious communists. The rest were students, academics or business people.

1980s–90s: a period of Eastern European immigration with refugees escaping war and political unrest in Romania and the former Yugoslavia. | 2000 onwards: economic migration from Eastern Europe facilitated by the enlargement of the European Union. |

Table 2
Historical immigrations to the UK, 19th century to present day

The geographical distribution of cultural groupings

- Across England and Wales, the proportion of the population from ethnic minorities rose from about 6% in 1991 to about 7.9% in 2001.
- The distribution is very uneven. Ethnic minorities are concentrated in major urban areas, particularly London, with approximately 1.8 million migrants.
- London has the highest proportion of people from minority ethnic groups and two London boroughs have a majority of people of Black and Asian origin. People classified as 'white' by the census made up just 39.4% of the population in Newham and 45.3% in Brent. Meanwhile, Vietnamese is the second language in Lewisham.
- There is a higher proportion of people of Pakistani origin in Yorkshire and Humberside (2.9%) and the West Midlands (2.9%).
- Some 2% of the population of England and Wales are of Indian origin, with Leicester having the highest proportion (25.7%).
- Bangladeshis formed 0.5% of the population of England and Wales, with the highest individual proportion in the London borough of Tower Hamlets (33.4%).

Issues related to multicultural societies

Religion: Migrants may wish to adhere to their own religious calendars and practices; this can cause friction with employers, colleagues and authorities.

Language: Migrants find it difficult to obtain employment and to integrate if they do not speak the host country language. This may also restrict educational opportunities.

Housing: Migrants are generally poor upon arrival. This, coupled with low wages, leads to a concentration of ethnic groups in the poorest housing of major cities.

Economic issues: Migrants may not have equal opportunities to obtain employment, and may be subject to discrimination, prejudice and racism. The perceived cost of providing for migrants may feed resentment and racial intolerance among members of the host population. Migrants tend to be welcome in times of economic growth, but during recessions they are accused of taking local jobs.

Healthcare: Concerns have been raised about poor immunization levels among migrants, although this has improved along with literacy rates and efforts to produce information leaflets in the target language.

This topic continues on the next page

A case study of multiculturalism: Bradford

The city of Bradford provides an excellent case study with regard to some of the issues faced by multicultural societies. A 'conflict timeline' could be identified, as follows:

- April 1976: 24 people arrested in the Manningham area of the city when Asian youths confronted a National Front march along with police.
- 1977 Birth of the Asian Youth Movement: This movement was committed to the support of families facing racist immigration laws and winning the rights of communities to defend themselves. One of their central slogans was 'Here to Stay, Here to Fight'.
- 1981 Trial of the Bradford 12: Twelve young Asians faced conspiracy charges for making petrol bombs to use against racists. They argued that they were acting in self-defence and won their case.
- 1981: Bradford Council helped set up and fund the Bradford Council of Mosques. This helped the council to strengthen the position of the more conservative religious leaders and to dampen down the more militant voices on the streets.
- Throughout the mid 1980s, the focus was on religious and cultural issues, for example, the demand for Muslim schools and for separate education for girls and a campaign for halal meat to be served at school.
- Late 1980s early 1990s: Asian business community increasingly divided.
- 7 July 2001 Bradford Riots: Two people were stabbed and 36 people were arrested during clashes between Asian and white youths. A total of 120 police officers were injured in the violence. Estimated damage: £7m.

Some would argue that, in Bradford, multiculturalism has resulted in greater segregation. The consequence of some of the above actions was to create divisions and tensions within and between different Asian communities. From the 1980s onwards, Muslims, Sikhs and Hindus began to live in different areas and attend different schools and institutions as new council-funded community organizations and youth centres were set up according to religious and ethnic affiliations. The Asian Youth Movement, the beacon of a united struggle against racism in the 1970s, split, torn apart by multicultural tensions. In the summer of 2001, the towns of Bradford, Burnley and Oldham descended into conflict.

Essential notes

For further information on Bradford and Oldham, see:

- The Cantle report (2001): www.oldham.gov.uk/cantle-review-final-report.pdf
- The Ouseley Report (2005) which recommended a 'people's programme' to bring harmony to the city of Bradford: www.sra.org.uk/documents/SRA/equality.../ouseley-report.pdf

Examiners' notes

This section of the specification does not require a case study, but it will certainly add to your understanding and help to add depth to your arguments.

Separatism within and/or across national boundaries

Reasons for **separatist pressure** include:

- Claims by people with a minority language or culture within a larger political area. **Example**: the Basques in northern Spain and south-west France.
- Economic imbalances between areas which are economically depressed and the wealthier core of the country. Such areas are often also geographically removed from the centres of power. **Example**: the Scottish National Party and Plaid Cymru campaign, respectively, for an independent Scotland and an independent Wales. This has been partly satisfied by the establishment of the Scottish Parliament and the Welsh Assembly and some devolution of decision-making powers.
- Claims by minority religious groupings. **Example**: the largely Christian population of southern Sudan is seeking autonomy or independence from the majority Muslim population of the north.
- A perception that exploitation of local resources by national government produces little economic gain for the region. **Example**: the Scottish National Party feels that the exploitation of North Sea oil and gas has done little to develop the economy of Scotland.
- State collapse. **Example**: Yugoslavia collapsed leading to the creation of the countries of Croatia, Slovenia, Bosnia-Herzegovina and FYR Macedonia.
- Strengthening of supranational bodies such as the EU – many nationalist groups feel that they have a better chance of economic development if they are independent. **Example**: Scotland believes it could thrive within the wider patronage of the EU by developing particular niche markets.

The consequences of separatist pressure may be peaceful or violent. There have been a wide range of actions and campaigns linked to demands for greater autonomy:

- The establishment and maintenance of societies and norms with clear cultural identities within a country. **Example**: the Bretons in France.
- The protection of indigenous languages through the media and education. **Example**: Welsh, Catalan, Manx.
- The growth of separate political parties and devolved power. **Example**: the Scottish and Welsh Nationalists.
- Civil disobedience. **Example**: the Orange Revolution in the Ukraine.
- Terrorist violence (widely employed by the Basque separatists – see case study on pp. 86–87).
- Civil war. **Example**: Tamil Tigers.

Examiners' notes

It is important to be able to give **examples** to support each of the types of separatist pressure listed here.

Essential notes

Autonomy: self-government

Separatist pressure: pressure by a group of people within one or more countries to achieve greater autonomy (ideally independence) from a central government from which they feel alienated in some way.

Essential notes

Those wishing to describe their ethnicity as Cornish were given their own code number (06) in the 2001 UK census for the first time. About 34 000 people in Cornwall and 3 500 people in the rest of the UK wrote on their census forms in 2001 that they considered their ethnic group to be Cornish.

The Jedi census phenomenon was a grassroots movement, organized via email, to encourage people to record their religion as Jedi – in fact, 390 127 people (almost 0.8%) in England and Wales did so, making Jedi the fourth largest reported religion in these countries, surpassing Sikhism, Judaism, and Buddhism.

This topic continues on the next two pages

Case study: the Basque separatist movement, Spain

Spain has a number of communities with strong nationalist movements based primarily on language differences. Distinctive languages are found in the regions of Galicia (Galician), Catalonia (Catalan) and the Basque country (Euskara). The following case study focuses on the Basque separatist movement.

Location: The Basque language, Euskara, is unique and distinctive, but the extent of the Basque country itself is subject to debate. There are four provinces in northern Spain that reflect Basque culture – Bizkaia, Gipuzkoa, Araba and Nafarroa – together with Lapurdi, Benafarroa and Zuberoa in SW France.

Fig 3
The Basque regions

Background: Franco, who led Spain until 1975, executed or imprisoned thousands of Basque nationalists. The use of the Euskara language on buildings/road signs or in publications was banned, and the teaching of the language was declared illegal. The culture and the language were suppressed for over 40 years.

Formation of ETA: In 1959 a national political organization, ETA (Euskadi ta Askatasuna), was formed. In the 1960s it declared war on the Spanish state. ETA has operated a violent campaign, targeting police, security forces and political and government figures and buildings. It has stated that its armed struggle will continue until it has achieved an independent Basque country, comprising both the Spanish and French provinces. In 1979, a government-held referendum in Spain resulted in massive support for autonomy of the Basques and a Basque parliament was created. However, continued attacks over the years alienated many Spaniards. Hundreds of thousands marched in Madrid in January 2000 to show their disapproval of ETA's violence.

Examiners' notes

Whichever case studies or examples you use, make sure that the detail you learn is specific, to provide the all-important sense of place. Hint: try covering up the name of the location in your practice answers – can you still tell where the case study is from? If so, you're on the right lines!

ETA is considered by Spain, France, the wider European Union and the USA to be a terrorist organization, with more than 800 killings attributed to the group. Its motto is *Bietan jarrai* (Keep up on both), referring to the two figures in the ETA symbol. The snake represents politics and secrecy, while the axe around which it is coiled represents armed struggle.

ETA's recent activities:

- March 2006: ETA declared a permanent ceasefire, stating that it would commit 'to promote a democratic process in the Basque country in order to build a new framework within which our rights as a people are recognized, and guarantee the opportunity to develop all political options in the future'.
- December 2006: A bomb in a car parked at Madrid airport left two people dead, and prompted the government to call off peace talks. The Victims of Terrorism Association held a demonstration in Madrid demanding that there should be no further negotiation with the terrorist group.
- 2007: Thousands of Basque citizens participated in a rally in Bilbao to protest against the decision of a Spanish court to order the detention of 46 members convicted of aiding ETA through a network of ostensibly legitimate social and political organizations.
- 29 July 2009: A car bomb exploded in the northern Spanish city of Burgos, injuring around 50 people – again, blamed on ETA.
- 16 October 2009: After the arrest by the Spanish government of several Basque politicians, a bus, several cars and three banks were attacked by petrol bombs in the northern cities of Bilbao and Ondarroa. The attacks were blamed on the *kale borroka*, youth groups related to ETA.
- 1 March 2010: Spanish judge Eloy Velasco accused the Venezuelan government of assisting ETA and Colombian rebel group Farc. Six ETA and seven Farc members were charged with terrorist plots, including a plan to assassinate Colombian president Alvaro Uribe.
- August 2010: Violence spread throughout the Basque country after a series of coordinated *kale borroka* attacks. Dozens of waste containers were set alight in the towns of Ondarroa, Zarautz, Azkoitia, Vitoria and Bilbao. Petrol bombs and other explosives also featured in the attacks, believed to have been orchestrated by ETA.
- 5 September 2010: ETA declared a new ceasefire – will this one be permanent?

Examiners' notes

This section of the specification does not strictly require a case study. However, as an essay could easily be set on this topic, you would be well advised to know one example in detail and perhaps two or three others to a lesser extent. Remember, the emphasis should be on the nature of the separatist cause, the reasons behind the demands and the consequences.

Essential notes

Kale borroka:
This roughly translates as 'street fighting' and refers to urban guerilla actions carried out by Basque nationalist youths. Tactics include attacking the offices of Spanish political parties, the property of people linked to these groups (burning cars and setting homes alight), attacking and destroying ATMs, bank offices, buses, and rioting at demonstrations, using molotov cocktails and burning waste containers.

The challenge of global poverty

Background

Around 1.4 billion people worldwide survive on less than $1.25 a day; in sub-Saharan Africa, this is nearly 50% of the population.

Nearly 15% of the world's population suffers from hunger, with 30% of the population undernourished in sub-Saharan Africa and 20% in South Asia.

As many as 10% of children in sub-Saharan Africa will die before they reach their first birthday.

A pregnant woman in sub-Saharan Africa runs a 10% risk of dying in childbirth, while in southern Asia 60% of women have to cope with childbirth without the assistance of a skilled healthcare professional.

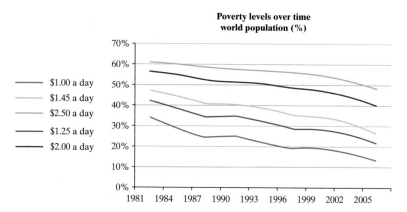

Fig 4
Change in poverty levels 1982–2006

World Bank indicators suggest that if we take the increased population between 1981 and 2005 into account, the poverty rate has actually fallen by about 25%. While this sounds encouraging, it masks regional variations, and in particular the impact of China's rapid economic development.

China's poverty rate fell from 85% to 15.9%, or by over 600 million people, accounting for the majority of the world's reduction in poverty. Excluding China, poverty fell by only around 10%.

Key terms

Poverty: The condition of not having the means to afford basic human needs such as clean water, nutrition, healthcare, education, clothing and shelter.

International poverty line: The poverty line (sometimes known as the poverty threshold) is the minimum level of income considered necessary to achieve an adequate standard of living in a given country. In the past, this was commonly held to be roughly $1 a day. However, in 2008, the World Bank came out with a revised figure of $1.25 at 2005 purchasing power parity (PPP).

Purchasing power parity: A technique whereby the purchasing power of different currencies is equalized for a given basket of goods. However, it can be very difficult to find comparable 'baskets' to compare purchasing power across countries.

Essential notes

One measure of PPP is the Big Mac Index, popularized by *The Economist*, which looks at the prices of a Big Mac burger in McDonald's restaurants in different countries. If a Big Mac costs $4 in the USA and £3 in the United Kingdom, the PPP exchange rate would be £3 for $4.

Human Development Index (HDI): A statistic used to rank countries by their level of 'human development' with countries separated into developed, developing and underdeveloped countries. This is based on the GDP per capita, life expectancy and education of the countries.

Physical Quality of Life Index (PQLI): This is an attempt to measure the quality of life or well-being of a country. The value is the average of three statistics: basic literacy rate, infant mortality, and life expectancy at age one, all equally weighted on a 0 to 100 scale.

Gross Domestic Product (GDP): The total value of goods and services produced by a country in a year, usually expressed per capita.

Gross National Product (GNP): As above *plus* all net income earned by that country and its population from overseas sources.

Millennium Development Goals (MDGs): A global action plan instigated by the UN to achieve eight anti-poverty goals by a 2015 target date.

Examiners' notes

'Key terms' or definitions need to be learned for the examination. They appear several times in this chapter and in the glossary.

Commonly used indicators of poverty			
Economic indicators	**Demographic indicators**	**Indicators**	**Composite indicators**
GDP per capita	Birth rate	% with access to safe drinking water	PQLI
GNP per capita	Death rate		HDI
% of population living on < $1.25 per day	Fertility rate	% children enrolled in primary school	
	Population growth rate	% urban population	
	Infant mortality rate	% population undernourished	
	Life expectancy	Adult literacy rate	
		Number of people per doctor	

Table 3
Commonly used indicators of poverty

Addressing poverty on a global scale

Fig 5
The UN's Millennium Development Goals. The Millennium Development Goals (MDGs) are eight goals to be achieved by 2015 that respond to the world's main development challenges

The eight goals are meant to act on the global causes of poverty as follows:

1. **Eradicate extreme poverty and hunger**
   - Reduce by half the proportion of people living on less than one US dollar a day
   - Reduce by half the proportion of people who suffer from hunger
2. **Achieve universal primary education**
   - Ensure that all boys and girls complete a full course of primary schooling
3. **Promote gender equality and empower women**
   - Eliminate gender disparity in primary and secondary education preferably by 2005, and at all levels by 2015
4. **Reduce child mortality**
   - Reduce the mortality rate among children under five by two thirds
5. **Improve maternal health**
   - Reduce by three quarters the maternal mortality ratio
6. **Combat HIV/AIDS, malaria and other diseases**
   - Halt and begin to reverse the spread of HIV/AIDS, malaria and other major diseases
7. **Ensure environmental sustainability**
   - Integrate the principles of sustainable development into country policies and programmes; reverse loss of environmental resources
   - Reduce by half the proportion of people without sustainable access to safe drinking water
   - Achieve significant improvement in the lives of at least 100 million slum dwellers, by 2020
8. **Develop a global partnership for development**
   - Develop further an open trading and financial system that is rule-based, predictable and non-discriminatory; includes a commitment to good governance, development and poverty reduction, both nationally and internationally.

Essential notes

There is much more to this final goal. Find out more at: www.un.org/millenniumgoals

Also worth investigating is the work of the MDG Africa Steering Group, as there is much concern that many African countries are 'off-track' in terms of achieving these goals. See: www.mdgafrica.org

World leaders came together in New York on 20 September 2010 for a high-level event to renew their commitment to achieving the MDGs by 2015 and to set out concrete plans and practical steps for action.

Addressing poverty on a global scale: 'Think global, act local'

It is now generally considered that 'grassroots' or 'bottom-up' small-scale projects tend to work better at raising living standards in poor areas. This is because there is much more local consultation with regard to specific needs and, indeed, use of local knowledge and expertise. The emphasis is on using **appropriate** or **intermediate technology** which is affordable, available locally and uses local skills and materials to encourage self-reliance and self-sufficiency.

Case study: Oxfam and appropriate technology

Oxfam works directly with many people all around the world, helping them to obtain better healthcare and education. See if you can match up their achievements with the relevant MDGs:

- In Zugdidi in rural Georgia, Oxfam supported local organizations as they set up basic, affordable healthcare in villages with many former refugees. They built and renovated low-cost clinics in 27 communities.
- Oxfam is working with people affected by HIV and AIDS. In Malawi, Oxfam trains and supports home-based carers: local volunteers who support the ill, elderly and orphaned in their communities.
- In India, Oxfam's HIV/AIDS programme aims to increase access to prevention, treatment, care and support for those infected and affected, as well as raising public awareness about the disease.
- In Sri Lanka, south Sudan and Liberia, Oxfam supports the rehabilitation of community health centres and the development of basic services, including medicine supplies.
- In Afghanistan, Oxfam helps to run special winter schools to help students catch up on schooling during the long, winter months. It is providing educational materials and training teachers, while continuing to lobby for aid to help Afghanistan invest in its education system.

Security and development

A key dilemma faced by charities and governments is summarized in the slogan: 'No development without security and no security without development.' It might seem obvious to suggest that development measures cannot be effective without security, while in turn, peace-keeping efforts will lead to nothing if there are no development prospects. Sustainable development, along with economic and social progress, is impossible without effective government and respect for human rights. Former UN Secretary-General Kofi Annan was certainly right when he stated that: 'Development policy is an investment in a secure future.'

In the 21st century, all countries must advance this cause by ensuring freedom from want, freedom from fear and freedom to live in dignity. In an increasingly interconnected world, progress in the areas of development, security and human rights must go hand in hand. There will be no development without security and no security without development.

Examiners' notes

Consider some of the case studies and examples used in this chapter or those you have already studied independently. How could you use them to address the issue: 'No development without security and no security without development'?

Unit 3, Geog 3: Contemporary geographical issues

This unit is tested by a 2½-hour examination. It consists of three sections, A and B and C. Section A comprises the three physical options; section B comprises the three human options; section C is the essay section.

You must choose one option from each of sections A and B. Each option is made up of structured questions. These are awarded marks based on the level of quality of the answer. The total mark for each option is 25.

Section C contains six essay questions, of which you choose one. Each question is based on one of the six options. This must be a different option from those you have chosen in sections A and B. These questions are synoptic in nature and require you to show your knowledge and understanding of different aspects of geography as well as any connections between these different aspects. The essays are also awarded marks based on levels. The total mark for the essay is 40.

Below you will find sample questions followed by sample average answers and sample strong answers, and also tinted boxes containing the comments and advice of examiners.

Plate tectonics and associated hazards

1 *Study figure 1, a photograph of the Sichuan area of China after a recent earthquake. Using figure 1 only, comment on the evidence that suggests that an earthquake has recently taken place.* [**7 marks**]

Fig 1

Average answer

From looking at figure 1 it is clear that the area in question in China has been severely affected by an earthquake. The region is prone to many high magnitude events of 7.0 or larger every year and this is partly due to the collision of the Eurasian plate with the Indian plate thus indicating a continental/continental convergence.

Much of this is both vague and irrelevant. The student is not answering the question yet

The evidence for a recent earthquake in the area is shown by the mountains in the background which suggests that a landslide has occurred due to the severe ground shaking caused by surface seismic waves. Furthermore, the area is also deserted with the only people present being soldiers who are probably part of a rescue team. This could suggest a high fatality rate or that the residents have tried to escape the area. Furthermore, figure 1 also indicates that some of the buildings have collapsed and walls and windows have fallen down which is another consequence of severe ground shaking and buildings being constructed on unconsolidated materials beneath. There also seems to be a high level of smoke in the area, which suggests that underground pipes could have been broken which have led to fires.

This answer contains a lot of irrelevant material, which, although it will not lose the candidate marks here, does use up valuable time. References to evidence are imprecise. This just merits a low Level 2: 5 marks.

Strong answer

It is clear that an earthquake has recently taken place for a number of reasons. In the middle of the picture a large building has partly collapsed and there are piles of rubble where the buildings once stood which implies that the earthquake tremors led them to collapse.

Here is a clear set of statements that take evidence from the photo and then make commentary on that evidence

In the background on the mountainside there is little vegetation and it is patchy. The rock appears fresh as if a landslide caused by an earthquake cleared much of the vegetation and the top layer of the rock away. In the background in the town there are columns of smoke which would indicate that some of the collapsed buildings have caught fire. This is probably due to the fact that gas pipes have broken and electricity cables might spark and set it alight. Also there appears to be extremely few people and animals present except for soldiers which could imply that the animals have fled and the people have been evacuated away from the area due to the earthquake. The soldiers are there to rescue all the people who are trapped in the rubble.

Overall, there has been a clear use of figure 1 and there are several comments about the nature of the evidence. There is some sophistication of description, and evidence of geographical thinking. Top Level 2: 7 marks.

2 *Describe the ways in which seismic waves and earthquakes are measured.* **[8 marks]**

Average answer

Seismic waves and earthquakes can firstly be measured by a seismograph which determines the strength of the earth's movements. These readings are displayed on a graph which shows the intensity and size of the movements. These readings can then be translated to a numerical scale called the Richter scale. The Richter scale is a logarithmic scale from 1 to 10 showing the size of an earthquake and associating it with the relative dangers of such an event. There is another scale which measures the intensity of an earthquake on a scale from 1 to 12, with 12 being about 8.5 on the Richter scale. The primary waves (P waves) of an earthquake can be measured and cause lateral movement of the earth's surface perpendicular to the wave's direction. Secondary waves (S waves) cause horizontal movement parallel to the wave's direction. Surface waves (L waves) are the slowest and cause both directions of movement. All these types of waves can be measured and used to predict the associated damage caused, with the numerical figure thus showing how seismic waves and earthquakes can be measured.

> There are brief references to two types of scale and to the use of technology. The section on types of earthquake waves is irrelevant. Level 1: 4 marks.

Strong answer

Seismic waves can be measured on a seismograph which records and locates the size of seismic waves during an earthquake event. Seismographs can be used to help predict an earthquake epicentre and they can also be used to help predict future earthquake events by determining which areas are most likely to suffer from structural damage, landslides and soil liquefaction. Seismographs measure the amplitude of the seismic waves by measuring the distance between movement of the instrument and the spring, which has inertia in it. The degree of movement between the mass and the rest of the instrument helps geologists and scientists to accurately measure the magnitude and size of seismic waves. Earthquakes are measured on a Richter scale which records the magnitude of the event. The Richter scale is logarithmic and so each unit represents a 10-fold increase in strength and a 30-fold increase in energy released. So a magnitude 7.1 earthquake is twice as big as a magnitude 6.9 event.

Moreover, earthquakes can be measured on a 12 point Mercalli scale which reflects the effects of an earthquake. The more severe the earthquake the more destructive the effects and the higher the score on the Mercalli scale. The scale, however, relies on individual interpretations of the effects of an event and not everyone will agree on its effects, e.g. degree of ground shaking. Finally, an earthquake can also be measured on the moment magnitude scale which is a more up-to-date way of measuring earthquakes by geologists. However the Richter scale is still used to show the size of an earthquake for the public and mass media.

> This account covers three different scales (seismographs, Richter and Mercalli scales), with commentary as well as some detail about the scales and technology used. The final sentences also recognize how others use this information, and are therefore topical. Top Level 2: 8 marks.

3 *With reference to two seismic events you have studied, compare the ways in which earthquakes and their impacts have been managed in contrasting areas of the world.* **[10 marks]**

Average answer

In 2010 seismic activity at the Atlantic and Caribbean plate boundary caused an earthquake on the island of Haiti. This earthquake was very strong and highly destructive. Haiti is considered one of the poorest nations on the planet and so naturally there was little infrastructure in terms of predicting the earthquake or even warning people of the imminent danger. This did not help in its management. The earthquake destroyed many buildings including homes, schools and hospitals, creating a huge gap in the social needs of the country. Foreign aid was desperately needed as the people could not fund the clear-up themselves. Massive investments from international governments and non-governmental organizations were called upon. This aid provided temporary shelter for displaced people, essential food and water needs and some medical support. Disease spread quickly, especially in urban areas where the amount of dead was too much for services to deal with, leading to bodies piling up in the streets. After the event management had to continue to be funded by the foreign means but for Haiti any preventative or protection schemes are not feasible. All the people were able to do was rebuild their lives from scratch.

In contrast, the 1995 earthquake in Kobe, Japan, ☞

saw a very different style of management. Japan is a highly developed and industrialized country with economic means to help itself without international intervention. Although the Kobe earthquake caused massive damage and many deaths the aftermath of the event was less demanding than that of Haiti. Japan had emergency provisions in place to reduce the effects, for example the able fire crews were able to deal with the fires which quickly sprung up over the city, especially in the traditional wooden housing. Furthermore, Japan had the economic funds to act on the knowledge learnt from the event. Fire breaks, gaps in urban areas to stop the spread of fires, were installed across the city, along with open spaces for people to get to safety away from potentially collapsing buildings. Investments in new building strategies led to earthquake proof/resistant buildings capable of withstanding seismic activity thereby protecting people.

Despite there being two clearly named earthquake events, there is a lot of generic material on impacts and management in this answer. Much of the information could apply to any earthquake-hit area. Also, despite the command word 'compare' and the phrase 'in contrasting areas' there is little if any comparison. Level 2: 5 marks.

Strong answer

The Great Hanshin earthquake or Kobe Japan earthquake took place on 17 January 1995 in the early hours of the morning. The focus was 16 km beneath the epicentre on Awaji Island, 20 km from Kobe. The tremors lasted 20 seconds. This earthquake was caused by the destructive plate margin where the heavier oceanic Pacific plate sank under the lighter continental plate, the Eurasian plate, causing a subduction zone which eventually caused the earthquake. It measured 7.2 on the Richter scale. ☞

In total, 6 434 people lost their lives due to the earthquake and many thousands lost their homes. Many of these people lost their lives due to the poor living conditions in their wooden houses with most having very heavy lead roofs with up to two tonnes of weight on their roofs. In response to this much stricter building laws were introduced so that not as many buildings would collapse if it were to happen again. Other measures implemented included installing weights on top of high-rise buildings to stop them swaying as much in future earthquakes and reinforcing buildings ☞

with steel girders, again for added support. All these ways of managing the impacts of the earthquake were quite easy for the Japanese government due to them being a developed country. However, a similar earthquake but in a different part of the world had much larger impacts upon the country.

The Gujarat earthquake in India in January 2001 was extremely devastating and one of the largest earthquakes in that region of the world for around 100 years. It caused 20 000 deaths and left around 1 million people homeless. Many died in the aftermath due to lack of resources, whereas in the Kobe earthquake the Japanese were much better equipped to manage the hazard. However, the Indian government should be commended for them sending in thousands of troops to help with the clear-up as well as food, medicine and tents for all the people affected, especially as all four hospitals in Bhuj collapsed.

After the Gujarat earthquake there were fears of the spreading of diseases such as cholera and typhoid but due to the swift response of the Indian government this did not happen. However, around 20 000 cattle died in the earthquake, which severely affected the local people, especially from an agricultural perspective. After the Kobe earthquake due to the country being ☞

more developed they were able to manage the impacts a little better than in India. Large sums of money were funnelled into research and other initiatives, such as installing rubber on to the underside of bridges so they are less susceptible to collapse because during the earthquake a lot of the freeways and train tracks collapsed.

Overall, both earthquakes were extremely devastating but Japan was much better at recovering from the impacts even though their total damage costs were around 20 times higher (Kobe – 100 billion dollars; Gujarat – 4/5 billion dollars). I believe this is due to Japan being much more developed and having better plans for when earthquakes strike. Afterwards the Japanese also introduced particular days of the year dedicated to earthquake safety and children at school have regular drills in how to act in an earthquake. This shows how well the impacts have been managed by the Japanese. The rebuilding of the destroyed port very quickly also shows how well managed the impacts were.

This is a well-developed and rounded answer, with good elaboration of the management strategies employed in two clearly identified and different seismic events. Comparison is full and extensive. Level 3: 10 marks.

The essay question

4 *Discuss the extent to which volcanic and seismic events are major factors towards proving that plate tectonics theory is valid.* [**40 marks**]

Strong answer

Plate tectonics is a relatively new theory and describes the large-scale motions of the earth's lithosphere, which is divided into major and minor plates which overlie the asthenosphere. Current research suggests the distribution of volcanoes and earthquakes is the foremost factor in understanding the mechanics of the theory, but it is important to also consider earlier evidence such as biological and geological evidence and palaeomagnetism. A brief introduction which recognizes the thrust of the question, but also recognizes that other pieces of evidence may have a role. The student appears to be setting out a framework for that which follows

Seismic activity at constructive plate boundaries supports the plate tectonics theory as it supports the idea that plates are dragged apart at ocean ridges and the results this can have. For example, at the Mid-Atlantic ridge the Eurasian and North American plates move apart at 2.5 cm a year. According to the plate tectonics theory, the plates are driven primarily by convection currents in the asthenosphere, due to residual heat and nuclear decay from the earth's core. Slab pull, the force of gravity acting on plates and driving them into the asthenosphere, may also play a role. Tensional forces cause the plates to move apart leading to downfaulting of sections of crust and transform faults (at the Mid-Atlantic ridge at an interval of roughly 55 km) due to sheering of the plate. These create several small shallow earthquakes as potential energy is built up along these fault lines and suddenly released as seismic waves. This paragraph provides detailed understanding of geological processes that seem to support the assertion in the question. The student is still in control of the task

Similarly, the formation of ocean ridges can cause volcanic events. These take the form of submarine shield volcanoes. They support the plate tectonics theory as they are thought to erupt through long fissures (hot plumes of magma rising from the core), ☞

occasionally breaking the surface of the water, due to isostasy, as at Surtsey in 1965. Iceland was also formed in a similar way. Under Iceland the mantle is some 200 degrees hotter than typical mantle so Icelandic crust reaches thicknesses in excess of 20 km. This paragraph is confusing. The student fails to explain how isostacy impacts on the processes referred to, and the last sentence is not clear in its meaning

In the 1960s Hess suggested that crust must be destroyed at the same rate as it is created as the earth is not getting any bigger. Seismic and volcanic activity at destructive plate boundaries help to explain this. At these plate boundaries the earthquakes have the strongest magnitudes, as subducting crust travels the fastest. Other earthquakes may occur as the crust is deformed, for example the deep focused 7.6 magnitude earthquake which occurred in Sumatra in September 2009. The first earthquake was thought to have resulted from deformation within the descending Indo-Australian plate, rather than from movement on the plate boundary itself. In terms of volcanic events at destructive plate boundaries, according to the plate tectonics theory, they are thought to be caused by hydration melting of the saturated subducting crust. Hydration melting is triggered by the release of water from hydrous materials such as amphiboles within the subducting plate, resulting in a silica-rich melt which rises to the surface. During this time the magma will undergo magmatic differentiation to become an acidic magma which will either crystallize below the surface to become a batholith or escape the surface as explosive volcanism. In terms of knowledge and understanding, this is an impressive paragraph. The student clearly understands aspects of tectonics theory well. The main issue that needs to be considered though, is whether it is being used, or applied, to address the question. The paragraph ends without the link being made, which is a pity

The pattern of earthquakes at conservative plate boundaries, like the San Andreas fault line in ☞

California, is explained by the plate tectonics theory. It suggests it is the build-up of potential energy between the sideways-moving plates that is released as seismic waves, such as the 6.9 Loma Prieta earthquake in October 1989. This is a clear statement of association, which could have been improved with more detail of the plates involved and the relative rates of movement of those plates

Sometimes, earthquakes can occur many miles away from a plate boundary, such as the Gujarat earthquake in 2001. The reactivation of old fault lines and pressures of distant collision plate boundaries are thought to be possible reasons for this intra-plate earthquake, according to the plate tectonics theory. Similarly, the eruption of Mount St Helens in 1980 did not follow the usual pattern of earthquakes. Plate tectonics suggests that the magnitude of the earthquake was caused by intrusion. Pressure of magma built up within the volcano and was said to give harmonic tremors. The earthquake caused the volcano's bulge and summit to slide way in a huge landslide. This is another situation where the student has left the reader struggling to understand what is being said. The student is assuming that the reader knows as much as he/she does, and is losing clarity as a result. Make sure you state exactly what you mean

Likewise, volcanoes can occur far from plate boundaries, particularly in the Pacific plate, for example Hawaii. These are referred to as hotspots and they are formed by a narrow mantle plume (the Kilauea plume reaches a depth of 450 miles). However, there is some disagreement over this and some geologists argue that in fact the plumes come from the upper mantle instead. And yet another case in point – the link to the question is clear, but the use of technical language is unclear

On the other hand, it is important to remember that there are other causes of earthquakes, which cannot be explained by plate tectonics theory. Britain, for example, seems to be experiencing many small earthquakes due to isostasy. The crust of the east and south of Britain is down-warping but is rising in the north because of the postglacial pressure release distortion. This can increase stresses on the local crust which can trigger earthquakes. Even human factors can be associated with earthquakes; for example in Newcastle, Australia, a 5.6 magnitude ☞

earthquake was triggered by the subsidence of old mine workings. It is also thought that the building of dams can cause earthquakes. Small earthquakes can occur after filling a dam, firstly because of the extra load due to the weight of the water; and secondly because water seeping down into faults can cause them to move, as liquid acts as a lubricant enabling faults to slide more easily. Another human-related cause of earthquakes occurs through harnessing geothermal energy. This can cause tremors up to a magnitude of 3. This is much better. This paragraph is clear, appropriate and relevant to the question set. The student is arguing that although seismic events can be evidence for plate tectonics theory, they can be caused by other factors and therefore their significance as evidence is reduced

There are other major factors which play a major role in the plate tectonics theory, even some of the earliest evidence. These ideas revolve around continental drift and seafloor spreading. Alfred Wegener formulated the idea of continental drift in 1912, arguing that the continents were once joined as a single land mass, Pangaea. Fossil evidence includes the fossil of the reptile mesosaurus which was found in both Brazil and southern Africa. This supports the view that tectonic plates move because it suggests the land masses were once joined and creatures lived in the whole combined area, but when the continents drifted the species were separated, living on different continents. They couldn't have swum across thousands of miles of ocean, so the plates must have moved. There is living evidence too: an earthworm living in both South America and South Africa. The last sentence does not really add to the answer, though the preceding section has now moved the answer into an area hinted at in the introduction. Other evidence is being examined

New evidence was discovered in the 1940s with seafloor mapping. Oceanographers found that ocean crust was being created at mid-Atlantic ridges – this supports the theory because it implies crust was being stretched apart. In the 1950s a parallel pattern of magnetism found in the basalt rocks either side of the ocean ridge was revealed. The iron particles in lava erupted on the ocean floor are aligned with the earth's magnetic field. As the lava solidifies these particles provide a permanent record of the earth's magnetic polarity at the time of eruption. This is referred to as palaeomagnetism and supports ☞

the concept of seafloor spreading because it ties in with the idea that new crust is being created and being moved apart. Fred Vines uses palaeomagnetism to support Hess's theories of seafloor spreading at the Atlantic mid-oceanic ridge. Another form of evidence is now examined – palaeomagnetism. Perhaps a little more elaboration of the parallel magnetic striping concept could have been given?

Therefore, volcanic and seismic events are factors when considering the plate tectonics theory, but perhaps not the foremost factor. Plate tectonics is one of the newest sciences and fresh evidence and research is being formulated and investigated every day. The relative importance of the different evidence has changed over time; first the main evidence was fossil evidence for continental drift, in the 1950s, palaeomagnetism became the principal factor and today volcanic and seismic events are thought to be the major factor towards proving the tectonic theory is valid. There are many parts still missing in terms of the mechanics of the theory due to there currently being no way of discovering the ways of the core of the earth. However, in the next few years the ☞

next piece in the plate tectonics theory could be found; it may be the last or there may still be many missing.

Speculating on what the future holds is not the best way to end this essay. There is sound and frequent evidence of thorough and detailed knowledge, with high levels of understanding, but perhaps not explanation. The essay is confusing in parts. Examples are used well throughout and there is evidence of an ability to synthesize a wide range of material to address the task. The relative roles of theory and evidence are interrelated well, there is balance, together with some assessment of 'extent to which'. The account is purposeful, and the writer is confident – perhaps too confident as clarity is lost on occasion.

Marks:

Knowledge: Level 4
Critical understanding: Level 4
Case studies: Level 4
Synopticity: Level 3
Quality of argument: Level 3
Overall Level 4: 32 marks

Weather and climate and associated hazards

1 *Study figure 1 which shows temperatures in London on a November day. Describe and comment on the extent of the urban heat island shown.* [**7 marks**]

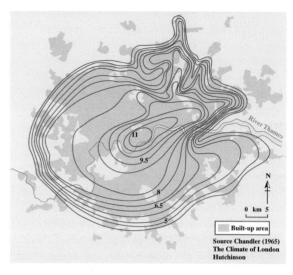

Fig 1

Average answer

An urban heat island is shown mainly concentrated in the centre in what seems to be the CBD (central business district) of London if the isobars are to be studied. The isobars show that a small area has a temperature of 11C which is a great difference compared to the temperature in the rural areas which is at less than 5C. This is the warmest part of London and so suggests to me it is the centre of the town. *This is clear description, but the commentary is weak*

Furthermore, this heat island effect extends to the east of my proposed CBD to an area of temperature 9C. This is still a significant temperature increase and suggests that the heat island effect is perhaps very prominent in London. Moreover, the river seems to be having a great effect on the heat island as it is cooler than surrounding urban areas. This suggests that the river area is a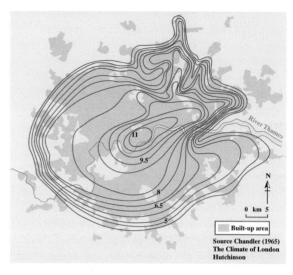

port full of industry which is producing this excess heat and releasing it to the urban environment. *A weak statement of commentary, which is also contradictory, as the river area shows lower temperatures*

Overall, I would state that the urban heat island effect is of a great extent in London as there are sudden temperature drops surrounding the urban areas as shown by the tightly packed isobars to the north and east. Also, most of the urban area is warmer than the rural temperatures showing the great extent of the heat island.

An answer that is sound on description, but the commentary is weak, and simplistic. The student recognizes the urban heat island effect at the outset, but does not use this to good effect. There is also incorrect use of terminology – isobars instead of isotherms. High Level 1: 4 marks.

Strong answer

The temperature distribution shown displays a clear difference in temperature between the central urban area of London and the surrounding area with maximum temperatures of 11C in the centre of the urban area and minimum temperatures of 4.5C on the outskirts and surrounding areas of the city. There is a steep decrease along the north and northwestern and northeastern edges of the urban area suggesting that there is a somewhat sharp contrast between urban setting and countryside. Good description here This steep decrease occurs because of the difference in materials between urban area and countryside. Those used in a city such as London include concrete, tarmac, brick and glass, which have different physical properties to the natural grass, soil and plants of the countryside. Urban materials absorb heat in the day (although little insolation here due the low angle of the sun) and release it slowly at night, meaning that night-time temperatures are higher than those of the surrounding areas. Good commentary A sharp contrast in materials, i.e. concrete to soil, would result in a decrease in temperature as shown here. Where the isotherms are further apart in the southwest, showing a gradual decrease in temperature, implies a more gradual transition between the urban area and countryside, perhaps an area of suburbs. Again, valid commentary on why the variations exist ☞

Where the river Thames occurs isotherms are fairly close together reflecting the cooling effect of the water but also how the heat lost from the urban area has also affected the inlet of water.

The urban heat island as shown by the isotherms is not a perfect circle nor does it perfectly correspond to the shape of the urban area. This demonstrates good understanding of the concept This is due to the distribution of urban structures and concentration of traffic where temperatures are highest (11C); this is most likely the location where traffic is most concentrated as emissions from vehicles heat up the atmosphere as do central heating systems in buildings. The 'valley' in the urban area just north of the river Thames and corresponding isotherms is most likely due to a park area, or an area of farmland, causing lower temperatures. Where isotherms are widely spaced suggests less densely populated or built-on areas, housing schemes and suburban areas.

This is an excellent answer providing good description of the temperature variations shown and offering good commentary on those variations. The student has recognized that an urban heat island is evident, and has offered commentary on its causes. There is strong evidence of geographical thinking. 7 marks.

2 Explain how urban design (street layout and architecture) affects winds. **[8 marks]**

Average answer

Urban structures and layouts can affect winds in many various ways. Firstly the uneven and rough surface of cities due to urban structures creates resistance, drag and friction to the wind. This affects both the direction and speed of the wind. Due to friction the winds tend to slow down   This is an effect   and are less powerful and fast than winds out in the rural areas without many urban structures nearby. Furthermore, the winds tend to follow the curvature of the buildings and many different winds coming from different directions may collide, therefore adding more resistance and slowing them further.   Repetition of idea

On the other hand, as well as making winds calmer buildings can also speed them up. Many high-rise buildings in a city may funnel winds. The wind cannot go through the building and so chooses the path of less resistance either to its left or its right. If this is repeated throughout the city some winds may combine their strength with others all being funnelled into the same direction.   Another effect   They can turn into gales with strength enough to knock a person off their feet or dislodge a shop sign.   This point is very simplistic, lacking in sophistication

Another effect the shape and layout of buildings has on winds is that it can funnel winds into small areas such as courtyards or car parks. The wind hits the floor of the open space but due to the enclosure of buildings on all four sides it has nowhere to go but upwards. This may result in the venture effect where mini tornadoes can be seen lifting up litter or dust in an anticlockwise direction.   This is an incorrect use of the term: the Venturi effect. Also the statement regarding eddying is too extreme

> There are some simple statements of how buildings influence air flows such as acting as windbreaks, creating turbulence and causing a 'canyon' effect. Level 1: 3 marks.

Better answer

Urban street layout and architecture affect winds in that friction from urban structures and tall buildings cause winds to lose energy   This is an effect   and to be diverted (respectively) ultimately leading to calm conditions and light winds. When moving through an urban area winds must pass through areas of skyscrapers, oddly shaped buildings and brick-built structures. Energy is lost when the movement of air encounters friction and hinders its ability to move as fast. Friction can be small scale, such as rough surfaces of brick buildings, or large scale, such as large tall structures which wind must go around. This results in a reduction of wind speeds in urban areas.   More description with some explanation but there is repetition here   Also, the more densely built-up an area is the less wind has an effect as the air must be diverted around many structures and can in fact be diverted round the area altogether. In terms of layout, parallel built streets, especially long narrow streets with buildings either side, can cause wind to be funnelled down these passages creating 'wind tunnels' where wind speeds increase greatly; this increases the average wind speed of an urban area. The same effect occurs with narrow passages/gaps between tall buildings, wind is forced through in a wind tunnel creating high speed winds.   This is not clear enough to be separated from channelling, which is described earlier   In urban areas with a less structured layout, e.g. wide streets with buildings of moderate height set out in an unparallel layout, wind is less affected; while speed is still reduced by friction, the occurrence of wind tunnels is less as the conditions are not conducive to such an event.

> The answer shows some appreciation of the effects of building spacing but there is little appreciation of the complexity of flows around buildings. It would be nice to see some reference to attempts by modern architects to control problem wind flows. Level 2: 5 marks.

3 *Discuss how pollution reduction policies have attempted to reduce the incidence of particulate pollution and photochemical smog in named urban areas.* [**10 marks**]

Average answer

Since the introduction of industry to the global community, the incidence of particulate pollution and photochemical smog has increased exponentially. Along with the increasing use of vehicles in the world, industry has been the catalyst for this increase: emissions from vehicles contain toxic particulates such as soot and unburnt hydrocarbons while industry fumes contain similar materials along with harmful chemicals which react in UV light to form smog. While this paragraph is contextual, it is not relevant to the question set

Pollution reduction policies such as the Clean Air Act passed by the European Union The Clean Air Act was passed in 1956, long before the EU in the 1980s were put in place with the intention of lessening pollution to the atmosphere. The act called for industry to stop using fossil fuels in particular, which when burnt released harmful pollutants such as sulphur dioxide. This act introduced the use of sulphur-free coal as an alternative. Inaccurate again: smokeless fuel was a response to the act However, this was much more expensive and led to an increase in the use of natural gas as a fuel which was also a 'cleaner' option than coal. The act introduced fines on companies who failed to meet emissions standards and incentives for those who did. In doing this the act was somewhat successful in reducing harmful particulate pollutants in the UK especially but also all over Europe. Some generic policies here, many of which go beyond urban areas, but Level 1 credit awarded nevertheless ☞

Other policies aimed at reducing pollution include the London Congestion Charge, introduced in the 2000s to reduce the volume of traffic entering the city centre and simultaneously reducing pollution in the process. The charge was intended to encourage commuters to utilize public transport instead of driving and adding pollutants to the atmosphere. This scheme has been somewhat successful due to the comprehensive public transport system that London possesses which is convenient for commuters. Emissions in the city and the incidence of particulate pollutants and photochemical smog have decreased since the introduction of the scheme but are still at dangerously high levels. A specific scheme, located, described and discussed Other policies have attempted to encourage industry away from city centres to reduce the accumulation of pollution. However, this simply expands the problem and spreads it out over a larger area – a futile solution in the long term.

In conclusion, pollution reduction policies have been inventive and initiative in attempting to lessen the incidence of particulate pollution but have failed to establish any large-scale (global) policies to achieve this. Global policies are not related to question

Two specific policies are discussed: the Congestion Charge and the Clean Air Act (albeit with some inaccuracy). Mid-Level 2: 6 marks.

Strong answer

In modern times particulate pollution and photochemical smog have become serious problems. The presence of dust in urban areas is 1 000% that in rural areas. Photochemical smog is made when sunlight hits both oxides of nitrogen and hydrocarbons. A vague yellow fog is created reducing visibility, leading to road accidents and being a health hazard, some stipulating it to be a cause of asthma and other breathing problems. The Greek city of Athens is famous for its photochemical smog. A general introduction, but no mention of reduction yet

There are many different pollution reduction policies that can be put into place. For example, legislature and legal policies can be enforced such as the Clean Air Act policy of 1956 of Britain. This act banned any form of untreated coal to be allowed to be burnt. Soot used to be a main particulate pollutant in 1950s Britain and smog a mixture of smoke and fog was common. By introducing the Clean Air Act within a matter of years smog Not sure which type of smog this is but a specific policy has been discussed here was no longer a problem and the air was measured to be of better quality.

Another example would be traffic-calming schemes and traffic reduction schemes, their aims to reduce the amount of vehicles on the road producing particulate pollutants and oxides of nitrogen. policy The congestion charge policy in London is a set fee which people must pay in order to enter the built-up capital city. Since the charge was introduced traffic has decreased ☞

enormously in London and the number of vehicles on its roads has fallen. Some assessment of success here Now put off by the expensive congestion charge many previous motorists prefer to walk through the city. Public transport systems have also been used to reduce such particulate pollution. The more people sharing public transport means less cars will be emitting particulate pollution. By upgrading the transport system or providing incentives such as cheap fares more people would be encouraged to use public transport. This has been shown to work in Doncaster where there have been many improvements in the bus service. A new airport-like bus depot is kept clean and stylish and a bus corridor A third policy, located and discussed from Doncaster town centre to Rossington gives buses priority at traffic lights and smooth bus-only lane journeys which can be quicker than driving.

Many of these schemes and policies have been put into action around the world and have ensured that particulate pollution and photochemical smog have been severely reduced.

This is a fully developed answer, with some elaboration and discussion of a variety of pollution reduction policies. Each one is clearly different from the others. There is more concentration on particulates than photochemical smog. It did not get to the top of Level 3 because at least one of the policies could have been expanded upon. Level 2: 9 marks.

Essay question

4 *'Storm events in the British Isles have a variety of impacts and lead to a range of responses to them'. Discuss this statement.* **[40 marks]**

Strong answer

The British Isles is known for its constantly changing and generally wet weather systems. 90% of the weather is caused by polar and tropical maritime air which create a cyclonic weather system. This typical weather of the British Isles is common. However, in some rare cases the British Isles has seen some strong storms and extreme weather conditions which are unusual in the British Isles climate system. An introduction that sets a scene for what is to come, and clearly matches the focus of the question

In 1987 there was a huge storm to the southwest and southeast of England which devastated buildings and left 12 people dead. The primary impacts of the storm were storm surges along the south coast, especially in Devon and Cornwall in which the sea's tide was said to have raised around 2 to 3 m at the storm's peak, meaning that sea water flooded people's homes costing thousands of pounds to individuals. The huge amount of precipitation from the storm (40 mm) meant that floods were caused especially along low-lying, flat farmland near Devon and areas near to large rivers or basins. Floods due to precipitation were the main causes of death during the storm. 10 of the 12 people who died, died as they were trapped in either a house, car or swept away by the sudden rush of water. Another direct impact of British Isles storms is the speed of the wind itself. During the 1987 storm wind speeds were recorded at a maximum of 112 mph. The wind blew over fences, blew slates off roofs and blew away any light objects which were not attached to the ground causing a secondary hazard to people who could potentially be hit by an object. Clear statements of impact that are applicable to the case study identified. A good range of impacts

It is not just on land, however, where people may be affected by the direct hazards of a major storm such as heavy wind, storm surges or floods. The Sea Empress, an oil transport ship for Mobil, ran aground and sank close to the Shetland Islands, northeast of Scotland, ☞

due to the unusually violent weather systems causing huge waves and strong winds which ran the Sea Empress aground. The huge amount of oil spilt meant that huge amounts of government money had to be invested in a clean-up operation as wildlife, beaches and scenery were all destroyed by the disaster. A confused case study here. The Sea Empress did spill oil, but off the southwest Wales coast; it was the Braer that ran aground in the Shetlands. However, the idea is correct – some sympathy here by examiners

Impacts of a storm such as the great storm of 1987 in the British Isles continue beyond the primary impacts such as strong winds and floods. The British society of the south was dramatically disrupted as winds knocked out power lines, windows were smashed, slates were flung off roofs, trees were toppled over, bins were blown down streets, leaving rubbish lying all over the community. Floods also meant that sewers overflowed, people's houses were left uninhabitable. All these factors meant that huge amounts of money had to be paid by insurance companies and the government had to spend money on clean-up operations. Generic statements of impact here, but there is some recognition of the roles of decision makers – synopticity

When you compare the impacts of storm events in the British Isles to other storm events from around the world, however, you see that there are much more severe storms in terms of direct and secondary impacts than that of the British Isles; for example, Hurricane Katrina in 2005 devastated the New Orleans city when a storm surge sent water above the levees, eventually causing them to break. Flood water engulfed the entire city of New Orleans which has been built on low-lying flat land close to the sea. Unlike the UK, in which housing is generally built to a very high standard, in New Orleans wooden shacks in which many of the population lived were blown down or demolished by the flood water. Similar secondary impacts affected New Orleans as affected the British Isles in 1987, however in New ☞

Orleans 2000 people died as opposed to 12 in Britain showing that the severity of the impacts in the USA was arguably greater than that of the British Isles.

Clear assessment of impact here – by comparing it to another part of the world with a greater scale of impact. An unexpected approach but a nice way to be synoptic?

In terms of responses to the storms of the British Isles, British people tend not to board up windows and evacuate areas such as areas like Florida do in response to a storm warning. Initial responses to storms in the British Isles are usually to keep inside homes or to be fatalistic in which there is no change to people's daily lives. I think people in Britain take this approach because they are used to constant cyclonic weather for a majority period of time and do not expect extreme weather conditions such as that of the hurricane season in the Gulf of Mexico. In areas such as Florida and New Orleans, whole areas of people have been evacuated by the government due to a hurricane warning, something which is unlikely to happen as an initial response to a storm in the British Isles. More assessment of responses here

Responses during the storm, however, may be very similar to that of people in other countries such as North America in areas such as Florida. A logical move into the second half of the question For example, many people will choose to stay in their home as it provides security and safety from heavy precipitation or strong winds. British people may also choose to put sandbags by their door to prevent flood water entering their house, a rare response in the British Isles; however, it is common in the American hurricane season. Storm chasers are people who take the opposite approach, however: these people go out of their homes and approach the storm itself through interest in meteorology. Some comparison of response here

Most of the responses to storms come after the storm has passed, and it's at this time where the evaluation of the response is critical, i.e. how good the response is can be linked to how severe the impacts of the storm were. In the great storm of 1987 the response was rapid. The emergency services were immediately deployed to help people trapped by flood water, clear up objects blocking transport routes, provide medical attention to ☞

wounded or even security measures – not allowing people to cross unsafe bridges. For example, huge floods in Cumbria in 2009 meant that people could not use a bridge which was the only way across a major river. Instead people had to divert 30 km around it. A reference to an additional case study – evidence of 'breadth' – which is good In 1987 people volunteered themselves to clean up rubbish and objects off streets and asked others if they were OK. Relatives offered places to stay to people whose homes had been destroyed by flood water and insurance companies were quick to respond to hundreds of claims. Some specific statements of response here

Although the response after the storm was rapid and good by the British Isles and the British government, it can be argued that if the 1987 storm and other storms alike were of the magnitude of hurricanes in the Gulf of Mexico or the Indian Ocean, that we would be completely unprepared and unable to respond effectively as we are not expecting storms of this nature. Hurricane Ivan in 2004 was an F5 hurricane, wind speeds of 150 mph which hit the Florida coastline killing 2000 people. The response by the Americans was planned, evacuating the area beforehand as they could see it coming, preparing food and medical supplies. However, it is the response after the hurricane which was extremely impressive. Although 2000 people died compared to only 12 in the 1987 storm, the American response was huge. FEMA spent $2 billion on rebuilding areas hit most and immediately provided tents and shelter including food and medicine to the homeless, preventing spread of disease and further death. Military helicopters were rushed to coastal areas ready to rescue anyone trapped by the sudden 3 m storm surge. Over 2000 fire crews were rushed to the area from other areas not hit by the storm. Again an unexpected direction in the answer, but made valid by constant reference back to the UK event. The student is clearly evaluating the UK response by comparing it with elsewhere

These responses greatly outweigh the British capability of responses, however it can be concluded that though the British response to storms may not be as high calibre as responses such as America, British storms are not as bad as other global storms such as in America like Hurricane Ivan and Hurricane Katrina. ☞

So, overall, British responses are feasible in that our magnitude of response is good enough for the magnitude of storms the British Isles receives.

A rounded answer, which does precisely what is asked of it. The total length of this answer is 1340 words, which is longer than average, but it does demonstrate strong evidence of knowledge and critical understanding. Most examples are developed, with some breadth, and a wide range of material (some of it unexpected) has been well synthesized. ☞

The roles of decision makers are clear. It is a logical, purposeful and confident essay.

Marks:

Knowledge: Level 4

Critical understanding: Level 4

Case studies: Level 4

Synopticity: Level 3

Quality of argument: Level 4

Overall Level 4: 34 marks

Ecosystems: change and challenge

1 *Study table 1. This shows a range of characteristics of the London plane, a tree that is often planted alongside roads in urban areas. Using information from table 1* **only**, *comment on the advantages and disadvantages of the London plane as a roadside tree.* [**7 marks**]

Characteristic	The London plane
Aesthetic value (is it attractive?)	A tall elegant tree. Summer canopy open giving pleasant dappled shade. Twisted branches with network of fine twigs and pendant fruits give pleasing winter silhouette. Flaking bark creates attractive colours on trunk
Does it make a mess?	Leaves, fruits and bark may need clearing from streets and pavements
Pollution tolerance	Very tolerant of air pollution. Hairs on young shoots and leaves help to trap particulate pollution. Mature leaves tough and glossy readily washed by rain. Flaking of pollution-blocked bark may also be important
Pests and diseases	Rarely affected by disease and pests (although some shoots killed each year by a fungus)
Soil conditions	Very tolerant of poor soil conditions including compacted soil (although some stunting of growth by road salt)
Space	Although in open sunny positions a vigorously growing tree it is very tolerant of pruning and root disturbance
Safety hazards	Open canopy offers little resistance to wind. Trees rarely blow over or shed branches. Fine hairs on young shoots, leaves and fruits may cause irritation and even allergies in some people
Microclimate	Produces light shade, so some cooling effect and increase in humidity. Will intercept some rain especially when in leaf
Biodiversity	Not a native therefore relatively few associated invertebrate species. Provides valuable resting sites for birds. Sufficient light below canopy to allow significant plant growth

Table 1

Average answer

By reading through the research it is obvious that the London plane tree is a good tree to have in urban areas. A tall elegant tree is good because it will look nice to the public and also it can be taller than the lorries which squeeze themselves through urban areas. And ...? It seems that it is a low maintenance tree as only leaves and bark need clearing away and some small cutting back. Although it can stand a large trim which will save time and money to the local council. A comment on an advantage

In my view, it is a bad idea to put any tree next to a road due to my friend no longer being with us thanks to a roadside tree. This is not relevant, because the table does not mention road safety issues But apart from the hazard of crashing I feel it will make the area look nice, provide shelter for wildlife and will still look good with the heightened level of pollution.

Although it is tempting, it is never a good idea to personalize an answer. There has been some use of the data, and there is tentative comment. Level 2: 5 marks.

Strong answer

The planting of the London plane tree alongside roads in urban areas has both advantages and disadvantages. One advantage of planting these trees is that they provide aesthetic value by showing attractive colours from flaking bark and a pleasant winter silhouette. Another advantage of planting the trees is that they help to reduce air pollution by trapping particulate pollution using the hairs from the young shoots. Also the trees are rarely affected by disease and pests, meaning they will grow and live for longer. This is a commentary on an advantage Despite this, fungal infections can kill some shoots which would provide further planting issues. Another advantage of planting the London plane tree would be that it would help to increase biodiversity by providing nesting habitats for birds and still allow sufficient light through the canopy for further plant growth. Another comment on an advantage

On the other hand, there are several disadvantages of planting the trees. Firstly, the mess made from fallen fruit and bark needs to be cleared, increasing maintenance costs. A comment on a disadvantage Also, with the vigorous growth of the tree and pruning tolerance little room is potentially left. Furthermore, the potential health hazards caused by the fine hairs on the leaves, young shoots and fruit from the trees can cause irritation of the skin. This would benefit from an idea about the scale of the health hazard

Overall, it would seem sensible to plant the trees as the benefits outweigh the costs.

The answer provides some commentary on the nature of the evidence. It shows some sophistication of description, and there is evidence of geographical thinking. Advantages and disadvantages have been addressed. Level 2: 7 marks.

2 *Urban ecosystems are constantly changing. Describe and explain two of these changes.* [**8 marks**]

Average answer

In urban ecosystems a number of changes have taken place for them to be as they are today. Firstly, grassland has been taken away and replaced by concrete infrastructure such as buildings and car parks. One change, though there is some debate as to whether the compete destruction of an ecosystem can be classed as a change – it no longer exists This undoubtedly affects the balance of the ecosystem which is made up of biotic and abiotic things interacting within an area. Where a field or park has been covered with concrete for a car park, nutrients are no longer being used and the most active part of that area's nutrient cycle would be leaching where nutrients are lost completely. Also, by replacing grass for concrete you are dramatically increasing surface runoff and drainage which has an effect on surrounding ecosystems too. Urban grasslands Needs to be more specific, such as school playing fields and parks play a large part in biodiversity. They house mice, rabbits, sometimes foxes, as well as flora such as dandelion and nettles. Removing these takes a chunk out of that area's biodiversity, a negative effect.

Secondly, people are constantly trying to improve their gardens' aesthetic qualities by introducing new species. Second change Some of these come from abroad and affect an ecosystem's balance as they may colonize an area being spread by wind and animals. This would benefit from some examples Also, within people's gardens they are changing the natural grass into a more uniform lawn, that is to say with the aid of chemicals they are changing the make-up of their grass and with it, the ecosystems surrounding it.

There are two examples but they are only loosely applicable to changes in urban areas. The detail is weak. Level 1: 4 marks.

Better answer

The colonization of wasteland has occurred in urban ecosystems. This occurs on wasteland where there is little human interference and where the conditions are suitable (enough moisture, surface roughness is high to hold seeds for further growth and where there is little slope so that water does not drain away too quickly). The colonization of a wasteland starts when lichen and mosses begin to grow as they can retain moisture and require few nutrients. These die off, they help to form a protosoil where hardy plants begin to colonize (in particular, the Japanese knotweed with windblown seeds). The further colonization continues and so more soil is formed which allows for the growth of young trees such as willow. This is a basic description of one change, but it lacks the depth required to access the higher marks

Another change that has taken place within urban ecosystems is the ecologies along routeways such as old train lines. A second change Colonization occurs here due to the lack of human interference (as many railway lines are fenced off). A similar method of colonization occurs here and is aided by the large amounts of light in such areas due to past steam engine train fires clearing the areas of the more dominant species. Plants that tend to colonize along routeways are the American willowherb and mugwort.

The answer has identified two sets of changes and there is some detail but to get to the higher parts of Level 2 the answer requires much greater depth of description and explanation. Level 2: 6 marks.

3 *With reference to an example you have studied, assess its success as an ecological conservation area.*
[10 marks]

Average answer

Ecological conservation areas are key to protecting an area's ecosystem, especially those which are fragile. The Serengeti National Park and Ngorongoro named example conservation area is an example of a fragile ecosystem. Despite climate change affecting ecosystems the world over it is safe to say that it is humans who have the greatest effect on plants and animals.

The Serengeti is a rolling grassland dotted with trees and shrubs. Nomadic herdsmen graze their cattle on this land under what is now strict supervision. In the past the animals were threatened by poachers killing mainly the now endangered African elephant. The lack of funds ☞

meant there was little the people of Tanzania, the home of the Serengeti, could do. In recent years, however, tourism has boomed and park rangers can now protect their land. This in itself is evidence of a conservation area's success as people from the world over want to visit it. This is only tentative assessment Another reason for its now proven success is that they have managed the area zonally and given power to the people of the land so they can act in a way that suits them and the park's unique nature the best.

> The description of the chosen area is specific, but there is only a tentative attempt at assessing its success. Level 2 awarded: 6 marks.

Strong answer

Dulwich Upper Wood is located in southeast London near the site of the old Crystal Palace. It is situated on a mixture of old woodland and previous Victorian gardens. The main aims of ecological conservation areas are to encourage wildlife back into a city, reduce maintenance costs in an area, increase biodiversity in an area and provide cheap use of derelict land. Although these statements could apply to any ecological conservation area the candidate is laying down the criteria upon which success is assessed for this named example To assess the success of this ecological conservation area these four aims must be addressed.

Dulwich Upper Wood has succeeded in fulfilling the first aim of ecological conservation areas as it contains huge amounts of wildlife which were not apparent before. With over 40 species of bird and animals such as foxes, mice and bats Evidence linked directly to the named example it is obvious that Dulwich Upper Wood has brought a significant amount of wildlife back into London.

Dulwich Upper Wood has also succeeded in reducing maintenance costs of an area. As the area is self-sustainable, very little input is needed apart from one employee to clean the toilet and collect litter once per day. This means that maintenance costs are reduced ☞

and considerably lower than previously when the area was prone to vandalism and graffiti by youngsters (which had to be cleaned and fixed).

Furthermore, biodiversity has significantly increased since the creation of Dulwich Upper Wood. The area contains over 150 species of plant including foxgloves and the yellow pimpernel. The wood also has separate areas containing various shrubs and small trees (including the willow and rowan). Considering the introduction of such fauna, it is clear that the biodiversity of the area has been increased.

Finally, Dulwich Upper Wood has definitely provided a cheap use for derelict land. With very low maintenance costs (£65 per week – £3 380 per year) the wood is obviously a cheap use of the land. In conclusion, Dulwich Upper Wood is clearly an extreme success as it has succeeded in all four of the aims of ecological conservation areas.

> This is a fully developed answer, with good elaboration and clear depth of detail of the chosen ecological conservation area, and to the assessment criteria given earlier. Assessment is explicit. Top Level 3 awarded: 10 marks.

The essay question

4 *'The characteristics of vegetation within ecosystems in the British Isles have changed over time. These changes are more the outcome of human factors than physical factors.' Discuss this statement.* [**40 marks**]

Strong answer

Since the end of the last ice age vegetation has developed in the British Isles to create the climatic climax vegetation known as the temperate deciduous woodland. However, very little of that remains today, largely due to clearance by humans, and also through the introduction of non-native species to these lands. What we see today is a combination of both physical and human factors, and this essay will seek to examine which has had the greater influence. Good solid introduction which recognizes the focus of the task, and sets the scene for what is to come

At first there would have been a natural succession known as a lithosere. In a lithosere bare rock is the starting point. First of all bacteria and algae colonize because they can survive with very little nutrients. They start to break down the rock. Pioneer species such as lichen then grow. They assist in water retention and this further weakens the rock. Lichens decay and mosses start to grow, further helping water retention. A thin soil starts to develop because of an increase in organic material and weakening of the rock. In protected water-retaining sites grasses and herbs may start to grow fed by the humus from the decay of the lichens and mosses. These grasses and herbs will die back and bacterial action will cause decay producing humus which increases soil fertility. This means soil depth increases and small shrubs such as fern and gorse start to grow. These shade out the grasses and herbs. Decay of these shrubs adds humus to the soil, improving the structure. This means that pioneer trees and large shrubs are able to grow, for example rowan and alder. Soon after, taller trees outgrow these, for example birch. Finally, as the soil now is very fertile, slower growing trees, e.g. oak and ash, start to outgrow these trees reaching heights of around 40 m. The climatic climax community is established – the temperate deciduous woodland. A strong section on the processes behind a lithosere, and the climatic climax. This is evidence of detailed and accurate knowledge ☞

This type of woodland could be found all over England and parts of Scotland and Wales. Because of geological and climatic conditions, there were some modifications to this climatic climax. For example, in limestone areas such as the Peak District ash was more common. In chalk areas, beech was more widespread. Further north, the natural vegetation was more coniferous. The Caledonian forest as it is sometimes known consisted of rowan, pine and birch, though on the damper and milder west coast oak was also common. In some small areas, other successions took place, such as a hydrosere where the wet environments created a climatic climax of alders and ferns and sometimes willow. Wider knowledge shown here – breadth – together with evidence of synopticity: the links to geology and climate variations

In the last 2000 years, however, from the time of the Romans there has been large-scale human interference. Some tree species were actually introduced by people: the Romans brought the sweet chestnut, whose nuts were ground to produce polenta, a staple food for Roman troops. In the Middle Ages, sycamores were introduced from France, and since then many other species have been introduced such as the rhododendrum, and the infamous Oxford ragwort which 'escaped' into the wild. The second element of the question – human factors – is being introduced here. Good evidence of a range of background knowledge, with a degree of maturity. The student is demonstrating confidence in the task, too

There has been large-scale clearance for farming, and for charcoal, across the entire country. Timber was chopped down across Scotland to build ships in Henry the Eighth's time, as well as being preserved by the same king for hunting purposes, for example the New Forest. Later, the demand for pit props and railway sleepers all created further need for deforestation. For centuries woodlands across the country were 'harvested' with techniques such as coppicing and pollarding being widespread. ☞

Vegetation successions have also been modified by human activity. One way in which this occurred is that humans wanted to create high quality grasslands for grazing for their cattle and/or sheep. One such area is Salisbury Plain. This meant the climax community of temperate deciduous woodland is changed to a plagioclimax of high quality grassland. First of all, humans would have removed the climax vegetation. They planted grass seedlings and used organic farmyard manure to stimulate the growth of grassland. A succession here would still occur but when the grass has grown, animals are let into the field and they eat the grassland. This prevents any more succession from taking place, no other species can become established, a plagioclimax has been created. This type of landscape is now being preserved in Salisbury Plain, even though the area is still used as a training area by the military. Today these herb-rich grasslands, covering 14000 ha, support 13 species of rare plants and 67 species of rare invertebrates. Since 1993 the training area has become an SSSI (Site of Special Scientific Interest) and, since 1994, an SPA (Special Protection Area) for birds. Good case study material to illustrate how human activities have created an ecosystem that some wish to now preserve. In this case human factors are more important than physical

Another way in which human modification can occur is in upland areas, such as the North York Moors where heather now dominates. This is because of human activity; heather would be present without humans but in much reduced quantities. The main human activity which has caused this to occur is the lighting of fires. Humans lit fires to clear areas for sheep grazing and grouse shooting. This meant much of the vegetation was destroyed. Humans lit fires to encourage new heather saplings to grow. This resulted in many other vegetation species dying out and not appearing in the uplands, as well as the tight grazing by sheep. However, fire-resistant species such as heather flourished. The fires were set at optimum times, this meant that much of the nutrients from the heather were returned to the soil and not lost in smoke. New shoots then grew which were ☞

perfect for grazing by sheep and grouse. More case study material of human activity. This one lacks a degree of clarity and accuracy, but a valid point is still being made

In the 20th and 21st centuries humans are now trying to protect unnaturally created landscapes. Snowdonia, or Eryri, as the Welsh call it, is a biologically impoverished landscape, with large areas overgrazed. It is one of a number of beautiful but spoilt landscapes. What can be done? Do we try to restore the biodiversity of at least parts of the area to a supposed natural state? This strategy has been called re-wilding, and may include the reintroduction of large carnivores, e.g. boars and wolves. This is conceivable in parts of the Scottish Highlands but Snowdonia's open spaces are too fragmented. More realistic is the prevention of grazing in the ancient woodlands on the slopes, and the enhancement of plant diversity in moorland areas by a reduction in sheep grazing. Here, an attractive plagioclimax landscape has been achieved but with a price paid in biodiversity. Another interesting case study which again makes the point that human activities have now a greater role in the creation of ecosystems

The South Downs have lost the ecologically diverse sheep pastures that dominated the scene until the 1950s – again a plagioclimax created over centuries. However, we find here landscapes where natural and human processes and features combine to create beauty. The patchwork of fields, meadows and woods can sustain moderate levels of biodiversity that could be enhanced with the right management. Again, there is a strong human dimension to the landscape. This is acknowledged by the government agency Natural England, which uses the term Natural Area to describe areas of countryside which can be identified by their unique combinations of physical attributes, wildlife, land use and culture – an outcome that is a combination of human and physical factors. The student is recognizing the role of decision makers in modern times in preserving landscapes created by a combination of physical and human factors. This is effectively rounding the essay off, with another useful case study

In conclusion, it is right to say that both physical and human factors have had a significant impact on changes to vegetation in the British Isles. The relative role of these has changed over time. ☞

For thousands of years, natural processes were to the fore – mainly because humans were not numerous. Within the last 2000 years, human factors have been more important, with distinctive landscapes being created. There is often talk of preserving the outcomes of natural processes in other parts of the world, such as rainforests. Here in the UK, conservationists try to protect both the remnants of natural processes, such as residual natural woodlands, and the products of human activity such as chalk grasslands. This illustrates that over time both physical and human factors have been important, but in differing proportions at different periods of history.

A relatively lengthy account (1310 words) which is strong in detail, critical understanding and ☞

the use of case studies. A high level of insight is exhibited, and there is some creativity in some of the ideas and examples quoted. The answer is focused on the task throughout, and is written with purpose, confidence and some flair. It is a balanced answer.

Marks:

Knowledge: Level 4

Critical understanding: Level 4

Case studies: Level 4

Synopticity: Level 4

Quality of argument: Level 4

Overall Level 4: 38 marks

World cities

1 *With reference to figure 1 and using your own knowledge, comment on the advantages of out-of-town locations for shopping centres.* [**7 marks**]

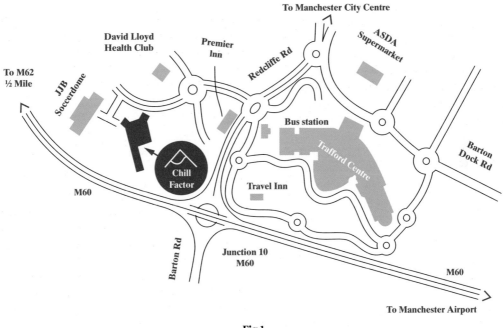

Fig 1
The location of the Trafford Centre

Average answer

The Trafford Centre seems to be highly accessible, close to junction 10 on the M60 which leads to the M62 in one direction and to Manchester airport in the other. This means that there are lots of potential customers within easy reach of the centre, ensuring profitability. Comment Land prices are also cheaper out of town which will have allowed developers the chance to provide many shops and services under one roof with landscaped surroundings and plenty of free parking, again attracting many customers. Comment In this particular case, the large amount of land available has enabled 👉

expansion to provide other facilities such as an Asda supermarket, a health club and a Soccerdrome. There are also two hotels encouraging customers to visit for a weekend or short break and spend even more of their money. Comment

This student begins to make basic comments that take the answer beyond what is obvious from the map provided. Although there is no real detailed reference to any particular case study, this still just merits Level 2: 5 marks.

Strong answer

One of the main advantages of an out-of-town location is the high degree of accessibility as compared with traditional town centre locations, which have become increasingly congested and notoriously difficult and expensive for customers and delivery vehicles to negotiate. Nice introduction, sets the scene Figure 1 shows that the Trafford Centre can be easily accessed via a slip road from a roundabout at junction 10 of the M60 and in fact a series of roundabouts circle the centre and associated leisure zone to promote free movement of traffic. Good comment The M60 even provides access to Manchester airport so that customers can come from far and wide to enjoy the vast range of facilities on offer. Good comment In fact, the Trafford Centre's catchment area is larger and more populous than any other regional shopping centre in the UK and comprises 5.3 million people within a 45-minute drive-time, with a total potential retail expenditure of £13 billion. Excellent use of case study

Another advantage is space; not only is there more room for both initial construction and expansion out of town, but land prices are also considerably cheaper. In some cases, brownfield sites have been utilized, as at Meadowhall which was built on the site of the old Hadley's steelworks. In such cases local governments are often willing to provide incentives for development, ☞

making the site even more attractive.

Good comment At the Trafford Centre developers have been able to utilize such space to offer 10 000 free car parking spaces, 230 stores, including the flagship Selfridges, a 20-screen cinema, Laser Quest, and an 18-lane bowling alley. The centre also boasts Europe's largest food court with seating for 1 600 and 60 restaurants, cafés and bars. Such facilities help to draw in customers and will keep them at the centre for a longer period of time, ensuring that they spend even more. One could even argue that the Trafford Centre in particular has taken the concept of out-of-town shopping a stage further, expanding to the west with its associated 'leisure village' shown in figure 1. This offers Chill Factor, with the UK's longest indoor real snow ski-slope, an Airkix indoor skydiving centre, a PlayGolf driving range, a Powerleague Soccer Dome and a David Lloyd Leisure Centre. Three Premier Inns ring the area making it an attractive option for mini-breaks, again ensuring maximum profits. Excellent use of case study and stimulus

This candidate makes good use of figure 1 and supplements the answer with excellent case study knowledge. Note that detailed reference to any out-of-town shopping centre would have gained similar credit here – it did not have to be the Trafford Centre. Level 2: 7 marks.

2 *Outline the causes of urbanization in an area you have studied.* [8 marks]

Average answer

Since the 1950s the most rapid growth in urbanization has occurred in LEDCs in South America, Africa and Asia. Between 1950 and 1990 the urban population living in LEDCs doubled, whereas in developed countries the increase was less than half. *A nice introduction, but not answering the question as yet – is it necessary?* One of the main causes of urbanization is rural to urban migration which is happening on a massive scale due to population pressure and lack of resources in rural areas (push factors). People living in rural areas are attracted to the cities because they believe that the standard of living in urban areas will be much better (pull factors). People hope for well-paid jobs or for greater opportunities to find casual or 'informal' work, ☞

better healthcare and education. *All basic Level 1 statements so far* For example, in Rio de Janeiro, the beaches of Copacabana and Ipanema are full of tourists – a potential source of income for those prepared to trawl the area offering drinks and snacks and souvenirs. Neighbouring industrial São Paulo also acts as a magnet with large factories such as Ford and Coca-Cola offering good, reliable wages and benefits such as free meals and medical care. *Two statements offering some relevant commentary which is correct for the areas identified, but not in any great depth*

This candidate just accesses Level 2 for the brief detail provided on Rio and São Paulo. 5 marks.

Strong answer

The term 'urbanization' refers to the increasing proportion of people inhabiting urban areas. This may occur as a result of rural-urban migration or as a result of high natural growth rates in urban areas, and often it is a combination of the two. According to the 2001 census of India, the population of Delhi was 13 782 976. In that year alone, the population had increased by 285 000 as a result of migration and by an additional 215 000 as a result of natural increase, making Delhi one of the fastest-growing cities in the world. The migrants themselves are typically in their reproductive years, so this contributes to the city's high growth rate. *Already Level 2 with correct figures for Delhi related to young age of migrants and their reproductive capacity* Pressure on the land in rural areas, such as parts of the neighbouring state of Uttar Pradesh, has resulted in the subdivision of holdings between family members. Farms have become fragmented and inefficient, resulting in declining yields and exacerbating poverty. *More Level 2 – sophisticated commentary, correct of area* By comparison, Delhi, as the administrative capital, appears to offer a host of employment opportunities. It is an important light industrial centre, with over 130 000 industrial units, producing everything from ☞

TVs to medicines, and seems to have emerged as the fashion capital of India, housing more than 60% of the design community. *More specific detail – more Level 2* In addition, Delhi's literacy rate of 81.8% compares favourably with a rate of 66% nationally, and 57% in Uttar Pradesh. Many of India's most respected universities and research institutes are here, such as the University of Delhi and the Jawarhalal Nehru University. There are also five medical colleges and eight engineering colleges. It is perhaps therefore not surprising that young people flock to the city in the hope of greater opportunities. *Even more specific detail – more Level 2* One could even argue that some of the excellent work undertaken by NGOs such as The Salaam Baalak Trust inadvertently serves to attract more migrants to the city as they are known to provide shelter, education and health care for street children. *Unusual angle – good geographical thinking – more Level 2*

This is an excellent answer identifying several different causes of urbanization with reference to a specific area: rural–urban migration in relation to employment opportunities and education, natural growth rates and the work of NGOs. Level 2: 8 marks.

3 *Urban regeneration may be achieved through partnership schemes between local and national government and the private sector. Describe one such partnership scheme and comment on its effectiveness in the area where it was established.* [**10 marks**]

Average answer

Partnership schemes such as City Challenge were based on a system of competitive bidding by local authorities, who had to develop imaginative plans involving the private sector and the local community to gain funding. By the end of 1993, over 30 City Challenge partnerships had been established and by the end of 1997 the government was able to claim that over 40 000 homes had been improved and 53 000 jobs had been created.

Good background information here, but not directly answering the question – despite the detail given, only Level 1 credit would be awarded at this point as the question asks for focus on a particular partnership scheme

The Hulme area of Manchester won one of these coveted City Challenge Awards resulting in significant redevelopment of the area between 1992 and 1997. Funding of £37.5m came from the government and over £60m from the private sector, bringing together ☞

organizations such as the Guinness Trust, Bellway Homes and Manchester City Council. Now at Level 2 – correct date, figures and partners identified The main achievement of the programme was the demolition of the 1960s crescents and deck-access housing which have been replaced with traditional two-storey homes and low-rise flats with attractive squares and courtyards. More Level 2 credit – correct of area Unlike urban development corporations, City Challenge schemes aim to be of more direct benefit to the residents of an area and this has certainly been the case in Hulme. Basic evaluation

The student only really begins to address the question in the second paragraph, but at this point offers some good background detail and a reasonable description of the change in housing provision. Detail on some of the other improvements would have helped this answer to move further up the level. Level 2: 6 marks.

Strong answer

Greenwich Millennium Village (GMV) has been developed on 121 hectares of brownfield site which used to house the largest gasworks in Europe. This in itself could be regarded as a success story as it has helped to transform an area which was once an eyesore into an innovative mixed-tenure modern housing estate which has already won more than 30 design awards. These include an 'excellent Eco Homes rating' for using and promoting environmentally friendly design and construction for housing. There is extensive use of glass, split bricks, corrugated panels, timber cladding and zinc sheeting – materials selected for their 'green' credentials. The development aims to cut primary energy use by 80% using low-energy building techniques and renewable energy technologies. ☞

Britain's first low-energy Sainsbury's is situated here and there is an integrated primary school and health centre with timber-clad buildings which reduce energy consumption by maximizing daylight and using more efficient systems for heating and air conditioning. Excellent case study detail – clearly suggests the scheme is a success. Already good Level 2 credit Other services include DIY and electrical retailers B&Q and Comet, along with a striking 14-screen UCI circular cinema complex with associated restaurants, all in the hope that GMV can be a self-contained and sustainable neighbourhood. Of course, in reality, most of the residents still work elsewhere and there is easy access to motorways, Canary Wharf and central London. More Level 2 commentary with a hint of assessment/ criticism English Partnerships ☞

initially had the overall responsibility for the project which has also involved two housing associations, Ujima and the Moat Housing Group, helping to provide housing for those on lower incomes. Well over £200m has already been spent in transforming the site and the project is due for completion in 2015 and will hopefully have provided in excess of 2950 homes. More Level 2 detail – named partners – implies success of scheme The importance of a natural environment was also recognized by English Partnerships throughout the development at Greenwich Peninsula. Three main areas of parkland have been created, including an ecology park, and extensive works have been carried out to improve the riverside environment. Not quite so detailed here, but still worthy of Level 2 – again, implies scheme is a success Nonetheless, the village has its critics – the housing is well beyond the pockets of the 👉

original inhabitants who feel disenfranchised and there is a distinct lack of community spirit as the newcomers commute into the City for work/shopping/entertainment, failing to bring new life to the area as was envisaged. No specific detail here, but a sophisticated comment which gives some negative evaluation and provides an effective conclusion

Overall, this is an excellent answer which describes the scheme in real detail on several counts – housing, services, natural environment. There is also clear positive evaluation in the first paragraph and clear criticism at the end. It would be hard to better this under examination conditions – the candidate builds a significant amount of Level 2 credit which would be lifted to an overall award of Level 3: 10 marks.

Essay question

4 Waste management has become a key issue in urban areas. Can sustainability ever be achieved in this context? **[40 marks]**

Strong answer

Sustainability is about meeting present needs without compromising the needs of future generations. Waste management in urban areas is increasingly becoming a key issue as the world is running out of space to dump its waste and the governments of various countries look to make a more sustainable world, both out of necessity and as a result of growing public pressure. This essay will look at the various techniques used in MEDCs and LEDCs and evaluate the extent to which these have achieved sustainability or not. A promising introduction which clearly sets out to answer the question asked, incorporates a definition of sustainability and promises to look at examples from across the development continuum

Waste management will vary considerably depending on the level of development of the country. This is mainly because MEDCs will already have successfully industrialized and the key issue for them is to achieve sustainability. For example, the UK industrialized in the late 19th century and early 20th century and waste management is now a big issue. 👉

One of the techniques which the UK has utilized has been through using Agenda 21, which sets out key sustainability targets for the UK's councils to meet. In order to meet these targets, the councils have pushed for more recycling in the UK. For example, in Swindon residents are each given a black box which they can fill with recyclable materials such as metals, bottles, paper and cardboard. This has been fairly successful in Swindon with recycling rates increasing from 18% to 24% between 2008 and 2009. This shows that the scheme is working but it is still a long way off from achieving sustainability because over 75% of the waste is still being dumped in landfill sites. Good case study detail here linked to Agenda 21, demonstrating knowledge and understanding

The EU has also set targets for the UK to meet its recycling rates; for example, by 2015 the EU wants the UK to be recycling at least 35% of its waste, and by 2020 the recycling rate increases to 50%. The UK is obviously still a long way off these targets with a national recycling rate of 20%. Therefore this suggests that the UK's waste management techniques are not currently successful 👉

in achieving sustainability. However, local schemes and initiatives are trying to change this. For example, in the Cotswold region Y-Use and IT Recycle have been set up to encourage businesses and locals to recycle and help achieve sustainability. Y-Use sells boxes and bags to the public who then fill their bags and boxes with recyclable waste. Y-Use then collects these bags from the public and takes them to the recycling centre where the materials are recycled. The business has been very successful and has helped to significantly increase the Cotswold's recycling rates to over 30%, significantly higher than the UK average. Excellent use of case study here

IT Recycle is an organization which helps businesses to recycle more. The company does this by taking away old computers, televisions, freezers etc. and taking them apart to find any materials that can be recycled or reused. Again, this company has been very successful, with estimates suggesting over 100 tonnes of material has been saved from landfill. These two examples show that although national schemes have only been partially successful the local initiatives have had a much greater impact and have certainly helped to achieve sustainability in these areas. However, although these have been successful, sustainability can only be truly achieved if all areas take part and cooperate with waste management. At the present time the UK is certainly not in this position because some areas are recycling, reducing and reusing at much lower rates than others. Again, excellent case study plus an element of synopticity; recycling rates are highly variable and dependent on the degree to which local councils and businesses are prepared to invest in them

Mumbai is an example of a city in an LEDC which is attempting to overcome the problems of waste management and achieve sustainability. Mumbai has a population of over 14 million and it has a large and extensive rubbish collection team. In the slum settlement of Dharavi, this is many people's main source of income. The people collect the rubbish around the city and sort it into recyclable materials. Materials such as glass, cardboard and metals are sold on to recycling companies. Other materials such as bricks and rubble are broken down and made into pottery by the locals. The pots are then sold on in the markets and the locals make a lot of money from doing it. This has a number of benefits and is certainly helping to achieve sustainability; ☞

the rubbish is being recycled and reused, which means less is being dumped in landfills. Locals have a source of employment and they have been able to run businesses which are benefiting the local economy.

Already building up into a very useful case study, demonstrating a significant degree of knowledge and understanding. We now also have some balance – an example from the developing world

However, although this initiative has been successful, Mumbai still has other problems regarding its waste management. For example, over 5 000 bins are collected from each day and dumped into landfill sites. Because Mumbai is still an LEDC it doesn't have the right infrastructure to cope with the waste and so the rubbish is just dumped on wasteland sites where no-one else lives. This is obviously preventing sustainability from being achieved because the country is not developed enough to deal with the masses of waste; currently over 500 tonnes of waste are dumped in the landfills each week. This also attracts vermin and the spread of diseases and this can significantly affect the health of the most vulnerable socioeconomic groups who do not have access to the right healthcare and medication. Nice link here – shows understanding of the wider picture Also, as Mumbai is rapidly running out of space to put its waste, this results in waste being left in the streets or dumped in the slums such as Dharavi – again, this is not sustainable as it is jeopardizing people's chance of a healthy future.

Although Mumbai does have people to collect recyclable material, the materials are not always being recycled, partly because of the poor infrastructure which means recycling plants are too far away. This often results in recyclable materials being dumped in wasteland sites because it is often cheaper and easier to do so, even though this is stopping sustainability from being achieved. This example again shows that the extent to which sustainability from waste management can be achieved is limited by the participation and willingness of a country or area to sustainably manage its waste. Element of synopticity here – the candidate recognizes that unless recycling is made easy, it won't happen!

Curitiba in Brazil is an example of a city which has arguably achieved sustainability to a large extent because of its innovative waste management strategy. With a population of over two million, this LEDC city ☞

has successfully ensured that recycling and sustainable waste management takes place in all districts. Curitiba has achieved this by ensuring everyone takes part in recycling. Residents are given two bags, one for organic materials and the other for non-organic materials. The bags are then collected each week by the rubbish collectors and sorted in the various warehouses across the city into paper, cardboard, metals etc. These materials are then taken to be recycled. This scheme has been enormously successful with over two thirds of rubbish recycled and over 500 tonnes of rubbish recycled each month. Curitiba has even received an award from the UN in 1990 for its recycling programme. Curitiba has been successful because it has made sure it has included everyone in the scheme, for example immigrants and disabled people can get a job in the recycling plants to help sort the materials. The most vulnerable socioeconomic groups have also been encouraged to reuse and recycle. For example, each week the people can trade their recyclable waste for free food, bus tickets or school exercise books; all of these items are extremely important for people and will make a real difference to their future. *Again, the link to improving people's lives here demonstrates a further synoptic link*

The Curitiba example suggests that sustainability through waste management can be achieved if the right approach is taken. The main recycling station has room to prepare people of different ages to learn and engage in the recycling processes. It has class and conference rooms, a playground, museum, and a collection of art and objects built with material collected. For every 15 trashed computers collected, one computer is made. Employees and children are learning to use them at the recycling station conference room. Thus, Curitiba has ensured that everyone wants to recycle, reuse and reduce because they know that they will benefit economically through jobs, environmentally through reduced waste and ☞

landfill sites and socially with free food, bus tickets and educational incentives. This also shows that the development of a country does not always matter because as an LEDC city, Curitiba has higher rates of recycling than most MEDC countries such as the UK and USA. *The case study is further developed here – significant detail is provided and again linked to sustainability*

In conclusion, the extent to which sustainability can be achieved through waste management will depend on a number of factors: for example, the infrastructure, the attitude of the community, the initiatives and the schemes available and the level of development of the country concerned. Arguably, the most important factors in this question will be the attitude of the community and the schemes used. As Curitiba has shown, the development of a country does not necessarily matter as long as you have the right schemes in place and people are willing to take part as a whole. *The conclusion rounds this essay off nicely, although it doesn't really offer anything extra/new – try to save something that little bit different for the conclusion if you can*

This candidate has certainly answered the question and in fact revisits and uses the words from the question in almost every paragraph – a good technique to use. The case studies are varied and well integrated – i.e. the candidate has applied his/her knowledge to the question asked. This is essential – many candidates show evidence of good case study knowledge but fail to apply it properly.

Marks:

Knowledge: Level 4
Critical understanding: Level 4
Case studies: Level 4
Synopticity: Level 4
Quality of argument: Level 4
Overall mid-Level 4: 36 marks

Development and globalization

1 *Study figure 1 which illustrates GDP growth and associated changes in the Chinese economy between 1978 and 2002. Using figure 1 and your own knowledge, comment on the trends shown.* [**7 marks**]

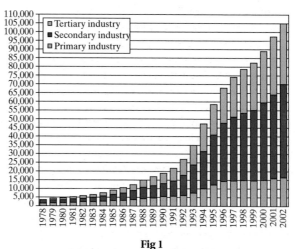

Fig 1
China's GDP by sector, 1978-2002 (in 100 millions of Yuan)

Average answer

China's GDP has grown dramatically from 4000 (100 millions of Yuan) in 1978 to 105000 (100 MY) in 2002. The fastest growth seems to have occurred in the early to mid-1990s, with GDP rising from 19000 (100 MY) in 1990 to almost 75000 (100 MY) by 1997. It seems that secondary industry has been mainly responsible for this rise, increasing from around 2500 (100MY) in 1978 to almost 60000 (100 MY) by 2002 – a 24-fold increase. This is not surprising as China is an NIC renowned for its cheap labour force. Tertiary industry has also grown in importance, hardly featuring at all in 1978, but contributing 35000 (100 MY) by 2002. Meanwhile, 👉

the GDP figures for primary industry also grew steadily, but at a much lower rate between 1978 and 1993, increasing more dramatically in the mid-1990s and then slowing again since.

This candidate probably feels that they have made good use of figure 1, but most of the data is simply lifted from the graph – this would only gain Level 1 credit. However, the statement that secondary industry has seen a '24-fold' increase in GDP value backed by the comment about cheap labour just allows this answer access to Level 2: 5 marks.

Strong answer

Figure I suggests that China has seen enormous economic growth between 1978 and 2002, with GDP tripling between 1978 and 1987, almost tripling again by 1993 and again by 2002. Growth seems to have been particularly fast in the early to mid-1990s, rising exponentially over this period. Good use of data here; the candidate must have thought this through very carefully to identify these trends Although there has clearly been growth in all sectors of the economy, it does seem that secondary (manufacturing) industry has led the way, contributing over half the GDP value from 1978 onwards. This is perhaps not surprising, as 1978 was an important year for the Chinese economy, marked by a change in government attitude and a decision to move from a centrally planned to a more market-orientated economy under the leadership of Deng Xiaoping. This opened China to TNCs wishing to use the country as an export platform and has made it a major competitor to other Asian NICs such as South Korea. Very clear commentary to back up the trends identified – already a good Level 2 answer

However, since the mid-1980s, it seems that the tertiary/ service industry has also begun to contribute significantly to the country's GDP. In fact, from 1986 onwards, ☞

about one third of GDP seems to emanate from this sector. This is partly a reflection of the huge growth in the Chinese software industry (the fastest growing of the service industries), stimulated largely by domestic demand. More and more newly 'wealthy' Chinese are acquiring personal computers and mobile phones that require software advances. In contrast, although showing some growth in terms of GDP value, the primary sector has been left behind. While it contributed almost half of China's GDP in 1978, this proportion had fallen to one seventh by 2002. China remains one of the world's leading producers (and consumers) of agricultural produce, producing about 30% more crops and livestock than the USA, but this sector has simply been outpaced by the performance of the other two.

This student uses the data very well, identifying a series of clear trends while avoiding the 'data waffle' which many candidates succumb to with these types of questions. Every point made is supported by a clear piece of evidence which displays very sound knowledge of the Chinese economy (a compulsory case study for this area of the specification). Level 2: 7 marks.

2 *Explain the recent growth of manufacturing industry in China.* [**8 marks**]

Average answer

Transnational companies locate their manufacturing plants in countries such as China so that they can take advantage of a cheaper labour force, and also less strict health and safety regulations. Companies such as Nike and Acer have done this. Although the wages would not be considered high in more economically developed countries, in China the wages are considered relatively high. This creates the positive multiplier effect meaning that more people have more money, therefore would be able to access more consumer markets and buy more goods. Thus quality of life increases, and more ☞

businesses move in. In China, this process was reinforced by the creation of SEZs which are special economic zones which make themselves more attractive to transnational companies by providing tax benefits and stronger infrastructure, along with more ideal locations for trading. Some even have electrical and gas facilities built in to further lower the costs of moving to the special economic zones.

> Most of this answer is at a basic level. However, the explanation of SEZs at the end of this answer just merits Level 2: 5 marks.

Strong answer

After the death of Chairman Mao in 1978 and under their new leader Deng Xiaoping, the Chinese wanted to bring an end to economic isolation. They had witnessed the success of the Asian Tigers as a result of foreign direct investment by TNCs and took the decision to move from a centrally planned to a more market-orientated economy. Initially there were strict rules and regulations on the businesses and enterprises set up, but these were soon relaxed. Government incentives attracted many TNCs who found they were able to maximize their production while taking advantage of a cheap workforce who were nevertheless relatively well educated. Although comments so far are general and could apply to many NICs, there is one statement regarding the Chinese move to a market economy that is relevant and specific, which accesses Level 2

The Chinese government designated certain areas as Special Economic Zones (SEZs) where companies were encouraged to locate in return for favourable taxation rules, the minimum of 'red tape' and improved infrastructure. These areas could offer wages 10% of those of the UK, so that an estimated £10 million could be saved if a firm chose to outsource 1 000 jobs to China.

Fourteen coastal cities, including Shanghai and ☞

Shenzen and three coastal regions, were designated 'open areas' for foreign investment. The coastal areas of China were chosen so that products could be easily shipped between China, the countries of South East Asia, and the rest of the world. This immediately led to FDI growth from $47 billion in the 1990s to $63 billion in the early 21st century. The growth of industries has led to China exporting 50% of the world's clothes and 20% of the world's shoes.

A big change also came in 2001 when China entered the World Trade Organization. This opened the country up to trade from all over the world as trade barriers were lowered and protectionist measures were reduced. As a result, the value of the country's exports increased from $30 billion in 2001 to $969 billion in 2007. However, it hasn't been without problems – unemployment is high in the inland provinces. The gap between rich and poor has increased and there are environmental concerns such as pollution. It is also a major concern that the Chinese economy is driven by exports.

> A very clear, logical answer which gives detail on three specific reasons for China's industrial growth – the change in political climate identified in the first paragraph, the creation of SEZs and the entry to the WTO, together with some commentary. Top Level 2: 8 marks.

3 *Discuss the importance of the role played by newly industrializing countries (NICs) in the global economy.* **[10 marks]**

Average answer

A newly industrialized country (NIC) is a developing country which underwent rapid industrialization in the 1960s. The first wave of NICs were the Asian Tigers which included Taiwan, South Korea, Hong Kong and Singapore. NICs generally attract transnational corporations such as big companies like Coca-Cola which operate in more than one country worldwide.

An introduction which sets the scene, but is not really addressing the question as yet

These Asian Tiger countries were originally targeted because they had a relatively high level of infrastructure, such as roads, rails and ports. Their workforce was also reasonably skilled, with cultures that revered achievement and education. They also had a good geographical location, government support was available and there weren't as strict laws on pollution, taxation and labour costs. These countries flourished, but soon they began to develop more and more until the price of manufacturing became too high and so the TNCs moved elsewhere, to countries such as India and China. *True, but again not really answering the question*

In general terms, NICs have a great impact upon the global economy, by attracting businesses to their country by offering incentives such as cheap land, less taxation and investments from the government. However, newly industrialized countries can have negative impacts.

Very general here – impacts on global economy are unclear

The impacts of transnational corporations' relocation to newly industrialized countries due to the attraction from various incentives can leave the host countries in economic turmoil. An example is in the UK where various companies which manufacture products such as cars and computers moved to NICs like Taiwan and China, costing thousands of UK jobs, which, in turn, put pressure on the local economy. ☞

This is better and hints at the impact, but the examples given still need to be more specific to access Level 2

The impact of NICs on the global economy as a whole, though, is good as it is able to keep prices low, so that these products can be sold to the consumer as cheaply as possible. Also, as NICs attract mainly TNCs, the TNCs have a great indirect impact on the global economy. The workforce receives very good training and so the population becomes more highly skilled. Then, even if the company does move away, the population still remains well trained. The TNC then moves to another area or country and trains that population. *A series of basic comments in this paragraph, which do not take the answer much further*

Another example of an NIC is Dubai, which was once a small pearl-exporting state in the United Arab Emirates in the 1930s. Then, in the 1960s, it discovered oil which led to a massive boom in the country's wealth and population. The population increased by 300% between 1965 and 1975 and it had a big impact on the global economy in many positive ways in the form of tourism, by creating the world's largest manmade harbour and building giant shopping malls, as well as creating the world's largest skyscrapers, free-standing hotels and sports stadiums. Also, due to its taxation laws, it attracted many large businesses to set up a centre there such as Microsoft, IBM, and Nokia. All of these positive impacts improved the global economy greatly, and it all started from a small newly industrialized country lying on the Persian Gulf, which discovered oil in the 1960s.

This is quite a long answer, but unfortunately it only really addresses the question in any detail in the final paragraph. The candidate accesses mid-Level 2 with the detail on Dubai, which is more specific and helps to substantiate the argument. 6 marks.

Strong answer

Newly industrialized countries have had a major impact on the global economy over a period of almost 50 years now. The four Asian Tigers of Taiwan, South Korea, Hong Kong and Singapore were the original NICs and started developing in the 1960s, experiencing meteoric economic growth which has forced the rest of the world to sit up and take notice. South Korea now has the world's 15th largest economy according to GDP and is classed as one of the world's richest nations; GNP per capita which was $100 in 1963 now exceeds $29 000 (2010). However, the South Korean economy is heavily dependent on international trade; witness the fact that in 2009, South Korea was the eighth largest exporter and tenth largest importer in the world. Clearly the country has a large impact and influence on the world economy.
A very strong introduction, already providing clear evidence of the importance of one of the Asian Tigers – already Level 2

In fact, all the Asian Tigers, together with second generation NICs such as Thailand, focused on export-driven industrialization rather than import substitution (tariffs on imported goods) in order to build up trade surpluses with industrialized nations such as the USA. At least to begin with, they largely ignored internal markets and developed through trade with MEDCs, flooding the world's markets with cheaply produced goods. They also encouraged investment by TNCs through the use of Export Processing Zones. For example, in Taiwan the government made use of EPZs to encourage TNCs as they could import goods free from tax as long as the products made were then exported. The value of exports is twice that of the value of imports through Taiwan's EPZs and thus the government has managed to increase the interconnectedness of Taiwan with the world and has encouraged flows of goods and capital within the world economy. *There is some clear detail here, with sophistication of understanding*

The NICs have caused businesses in MEDCs to become more efficient in order to compete and are now moving towards high-tech industries in order to add more value to products and to satisfy their own internal markets. Second generation NICs such as China and India have an impact on the global economy because of their rate of economic growth. China's economy grew by 9% last year and it is set to become the largest economy in ☞

the world by 2030. The consumption of products by China is vast which also has an impact on the global economy as it provides a massive economy of scale with its huge potential market. Chinese people are acquiring a higher purchasing power as 1% cross the poverty line each year to increase the number of wealthy middle class who can purchase products from the huge numbers of TNCs now investing in China. NICs allow TNCs to take advantage of huge cheap labour forces and have contributed to a buoyant retail sector in many MEDCs as cheap consumer goods are produced by NICs and are sold on the world market. For example, China can produce cheaper toys than the EU even when EU tariffs are applied to them. *More Level 2 credit here, although not quite so well stated. If the answer finished here it would certainly merit the top of Level 2*

However, the Asian Tigers also had a major negative impact on the global economy in the late 1990s through the Asian financial crisis, whereby a market crash was initiated because their stocks and shares were overloaded. The crisis started in Thailand with the financial collapse of the Thai baht and a burden of foreign debt that made the country effectively bankrupt. The crisis spread to most of Southeast Asia and Japan and other global markets consequently experienced some turbulence. *The other side of the argument – sophisticated, detailed and well stated. This answer now moves to Level 3*

Consumption by NICs has also had a detrimental effect on many of the world's least developed countries as they can no longer afford to purchase goods as prices have risen due to NIC demand, e.g. steel for the 2008 Beijing Olympics. While it is true that NICs have begun to close the developmental gap between MEDCs and LEDCs, it may well be that for the very poorest countries of the world such as Burkina Faso and Niger, the gap may continue to widen and they will remain stuck at the bottom of the development continuum.

This answer ends with an interesting and relevant summative comment demonstrating synopticity and certainly merits full marks. It is clear and logical, uses quite sophisticated terminology and provides a great deal of evidence to demonstrate both the positive and negative impacts of NICs on the global economy. Level 3: 10 marks.

Essay question

4 'Aid will always be necessary for countries at very low levels of economic development'. To what extent do you agree with this view? [**40 marks**]

Strong answer

Countries at very low levels of economic development (LDCs) are the poorest in the world and include countries such as Bangladesh, Sierra Leone, Malawi and Haiti. These countries have huge problems, reflected in their GDP being less than US$800 a year on average. The problems they face are numerous and include challenges such as war, famine, natural disasters, very poor quality of life indicators and of course a huge burden of debt. It could therefore be argued that these challenges cannot possibly be overcome by these countries alone and that international assistance through aid programmes is a necessity. *A sound introduction displaying reasonable knowledge and understanding of the challenges faced by such countries and directly addressing the question*

Aid may be short term with the aim of responding to sudden catastrophic events such as the Indonesian tsunami of 2004 or the Haiti earthquake of 2010. Events such as these highlight how dependent LDCs are on international aid. The scale of the Haiti earthquake was too much for this small island nation, the poorest in the western hemisphere, to cope with alone. The aid effort on the ground was spearheaded by Médecins Sans Frontières who have provided medical care to more than 92 000 patients and performed nearly 5000 operations since the earthquake struck. Surely no one could argue against such humanitarian aid. *Building a strong argument in favour of aid, backed by detail about the relief effort in Haiti*

However, aid may also be given long term through the development of projects, such as 'bottom up' or 'grass roots' schemes that are often stimulated by NGOs (non-governmental organizations such as the charity Oxfam). Such aid schemes are thought to better target the people who need aid the most, such as the PATH birth kits to assist in the delivery of babies or the WASH initiative developed by WaterAid in Burkina Faso. The former has helped develop 'clean birth' kits in Bangladesh, Egypt and Nepal over the past decade. ☞

Most kits contain a small bar of soap for washing hands, a plastic sheet to serve as the delivery surface, clean string for tying the umbilical cord, a new razor blade for cutting the cord, and pictorial instructions that illustrate the sequence of delivery events and hand washing. Aid such as this can be viewed as 'appropriate technology' – low cost, simple to use and sustainable – and it may go a long way to reduce the frightening figure of some 1600 women dying every day due to pregnancy and childbirth complications. It would be very difficult to argue against the validity and worth of aid schemes such as this. *Excellent detail on the PATH initiative – did the candidate forget to go back and talk about WASH? It hardly matters – the argument is again very convincing*

Moreover, aid can be given in three other ways. Multilateral aid comes from several different countries – often through international agencies such as the World Bank, whereas bilateral aid is aid given directly from one country to another, in the form of money, goods or services. The latter form of aid often unfortunately comes with 'strings attached' and may tie LDCs into agreements which ultimately put them at a disadvantage. For instance, US policy dictates that much foreign aid should be spent on costly imported medicines, weapons, agricultural produce or manufactured goods. Some recipients are even concerned that aid has only been forthcoming when they have agreed to join the alliance against terrorism. *The first hint that aid may have a downside – no specific example is given, but the quality of argument is quite sophisticated and there is a clear element of synopticity* Finally, celebrity philanthropy, through events such as Live Aid with Bob Geldof, could also be regarded as a new way of giving aid. Despite the pros and cons, it is this diversity of aid that is often viewed as its strength, as it can be provided in both humanitarian and monetary terms, although others would argue that aid is in fact damaging to the growth of LDCs as they can become aid dependent.

One major problem for LDCs is of course their ☞

debt burden. The debt is often so great that they have little hope of being able to repay it without some form of international aid. At the G8 summit in 2005, the world's eight most powerful nations set up an initiative to eradicate the debt burden for the world's Heavily Indebted Poor Countries (HIPCs) under the Multilateral Debt Relief Initiative (MDRI). The countries did have to ensure that they met certain criteria, such as having suitable plans in place to sustain economic growth, but by 2006, the G8 summit, pressured by the Live 8 concert and philanthropists such as Bob Geldof and Bono, had eradicated 17 countries' debt, allowing them a fresh start. This provision was used in Malawi, which ranks 160th out of 182 countries in terms of HDI (Human Development Index) and has a very low GDP of US $884 per capita. These figures, combined with agricultural problems brought about by drought, allowed Malawi to be relieved of £2 billion of debt under the MDRI. With this assistance from the rest of the world Malawi is able to spend more money on improving services and gaining preventative treatment for HIV/AIDS which is a prevalent issue. In this way, it is hoped that aid can help to initiate economic growth. Excellent breadth and depth of knowledge displayed here and the final sentence again hints at the necessity for aid

However, another huge challenge for LDCs is their inability to trade to advantage with other countries. The trade versus aid debate among economists and humanitarians is hotly contested. The three prerequisites for the stimulation of trade in an LDC seem to be that it operates a western-style democracy, it has resources that it can harness or at least learn to harness, and that it can sustain nationalism as opposed to protectionism. For many economists, trade is viewed as the most successful engine for growth and the experience of the NICs and BRICs would seem to support this view. China, once a centrally planned economy, has become more market-orientated since a change of government in 1978. The country created 'Special Economic Zones' (SEZs) where investors (often TNCs) could set up both plants and factories. These were based in fourteen coastal cities such as Shanghai and the combination of this accessible location, together with cheap, well-educated labour and considerable tax incentives, meant that much FDI was drawn to the area. This investment in trade has allowed China's economy to grow by approximately 👉

10% a year and seems to suggest that perhaps, after all, aid is not necessary. However, even with this example, foreign direct investment from the more developed countries was required to kick-start the process. Clever use of this case study to illustrate the point that aid is not the only way to achieve development – unless, as the candidate neatly suggests, FDI is in itself a form of aid?

Thus it is evident that although trade may eventually allow countries to become more independent, in the case of countries with very low levels of economic development it may still require initial assistance from the rest of the world. The trade solution also has other flaws for LDCs. Many critics believe that with diseases such as HIV/AIDS at near epidemic proportions, it may be impossible for these countries to flourish. This disease strikes at the heart of the economically active population and seriously undermines a country's ability to progress. It is also by no means certain that countries struggling at the bottom of the development continuum will ever be able to compete with the more developed countries. Many do not have the advantage of China's huge natural resource base or India's huge pool of cheap, but well-educated and largely English-speaking workforce. Even for those that do possess a wealth of natural resources, lack of technology may mean that these resources cannot be harnessed, or can only be harnessed with assistance from predatory TNCs. And as the development of new technologies proceeds apace, it is uncertain as to whether the LDCs will ever be able to invest in and catch up with such mandatory developments which may preclude them from being able to compete on an even playing field. A sophisticated level of understanding/synopticity is shown in this paragraph, which does link back to the theme of the question

In conclusion, aid may well be necessary to assist the world's least developed countries which have to face multiple challenges in terms of political instability, war, drought and the battle against AIDS/HIV. However, it can also create high levels of inflation (if the aid is used to fund the current account budget deficit) and it can lead to aid dependency. Aid, historically, has failed to create sustainable growth or development and food aid in particular may depress prices in domestic markets, so it must be carefully targeted. While no one would 👉

deny the need for aid on humanitarian grounds after major disasters such as the Haitian earthquake, it is the responsibility of those with influence in the more developed world to ensure that aid is appropriate, reaches the people who need it most, and does not have an adverse effect on local economies. Meanwhile, even the best thought-out aid schemes should not be seen as a substitute for fair trade. Yes, aid is necessary and probably always will be, but it cannot do the job alone.

The candidate has answered the question comprehensively and in large part has agreed that aid will always be necessary. He/she revisits and uses the words from the question at several points, emphasizing the fact that the argument is on track. It is certainly a good idea to do this. The case studies and examples are varied and ☞

well integrated, showing breadth and depth of knowledge along with a high degree of understanding and synopticity. It is also well written and sophisticated. The essay could be improved still further with a case study to show where aid has led to dependency or where tied aid has been a problem – this is hinted at, but not developed.

Marks:

Knowledge: Level 4 (high)

Understanding: Level 4 (high)

Case studies: Level 4 (low)

Synopticity: Level 4

Quality of Argument: Level 4

Overall, high Level 4: 37 marks

Contemporary conflicts and challenges

1 *Study figure 1, which shows the percentage of the world's population who were undernourished in 2006. Using figure 1 and your own knowledge, comment on the pattern shown.* [**7 marks**]

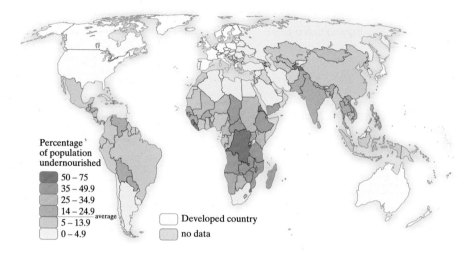

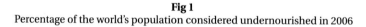

Fig 1
Percentage of the world's population considered undernourished in 2006

Average answer
Africa has the highest rates of undernourishment, with many countries having a rate of 35% and a similar number having rates between 20 and 33%. These countries are all in the 'poor south'. India also has a significant problem with 20–33% undernourished, but most of the rest of the developing world has achieved lower rates of between 5 and 19%, e.g. Brazil. The rich countries of the 'developed north' have the lowest rates, of less than 2.5%, although there is also a band in 👉

between across North Africa and the Middle East with rates of 2.5–4%.

This candidate has made reasonably good use of figure 1 in terms of identifying patterns, but the data is simply lifted from the map and would only gain Level 1 credit. Commentary is limited to statements about the 'rich north' and the 'poor south', but this idea is not developed in any way. Level 1: 4 marks.

Strong answer

The greatest problem with regard to undernourishment is clearly in the African continent and particularly sub-Saharan Africa where several countries have up to 35% of their populations undernourished. Examples include Angola, Zambia and Mozambique along with the island of Madagascar. This is perhaps not surprising in view of the fact that 50% of the population in sub-Saharan Africa have to exist on less than £1.25 a day and that they all have extremely low HDI rankings. *Some commentary here to back up the statistics* To the north of this zone several countries stretch in a band across northern Africa and exhibit high rates, of 20–34%, while only two countries in North Africa manage figures less than 4%, on a par with the more developed world. However, Africa is not completely alone and there are pockets of severe undernourishment (35%) elsewhere, such as in Afghanistan and North Korea, both countries associated with severe conflict situations which have had a clear impact on the health and nutrition of their populations. *More commentary to back up the pattern identified – already a good Level 2 answer*

The Indian subcontinent also figures prominently, ☞

with between 20 and 34% undernourished in India, Pakistan and Bangladesh, along with several countries in SE Asia, Central America and one in South America. The pattern in fact seems to reflect the development continuum, with many of the leading NICs such as Brazil, Mexico and China having much lower rates of undernourishment (5–19%) and some of the oil-rich nations achieving lower rates again (Saudi Arabia and Egypt for example, with between 2.5 and 4%). The lowest rates of all are found in the most developed nations (less than 2.5% in USA/Canada/Australia etc.) where food security is not such an issue.

This student uses the data very well, identifying some clear patterns and demonstrating excellent knowledge in terms of geographical location. Note that this is not a requirement – it is possible to identify patterns and distributions without reference to particular countries. The main point is that the candidate is able to back up the statements made with very sound knowledge of the challenges facing some of these countries. Level 2: 7 marks.

2 *What is meant by the term 'separatist pressure' and under what circumstances might this emerge?*
[8 marks]

Average answer

Language and culture in themselves provide strong factors which a group may wish to maintain. In France for example, the Bretons are particularly proud of their cultural heritage, which dates back many centuries. Separatism here occurs on a minor scale as the group actively maintain and upkeep their culture, passively contesting the 'majority' French culture. In the UK, belief in cultural heritage also affects the inhabitants of Cornwall, their history relating the people closely to other Celtic peoples such as the Bretons. Just enough detail to access Level 2 here

A divide in political ideology can be a particularly powerful factor in encouraging separatism. ☞

In Spain, ETA was set up in 1959 in response to their perception that the culture of the Basque region was being eroded by Franco's regime. Just enough detail to access Level 2

Finally, the collapse of large autonomous regimes may cause a variety of states or peoples to uptake separatist ideals. At the fall of the USSR many states became increasingly nationalistic, such as the Ukraine for example, which achieved independence in 1991. Not quite so convincing here

Although there is no actual definition of separatism, the meaning is implied through the examples given. Level 2: 6 marks.

Strong answer

Separatist pressure occurs when a group of people within one or more countries wishes to achieve greater autonomy (ideally, independence) from a central government from which they feel alienated. This may happen for a variety of reasons. For example, in the case of the Basques of northern Spain and southwest France, the people have a unique and distinctive language and culture and have campaigned for a separate state for many years. Similarly, the largely Christian population of southern Sudan is fighting for independence against the majority Muslim population of the north. State collapse may also lead to separatist pressure as in the case of the break-up of the former Yugoslavia, leading to the creation of the countries of Croatia, Slovenia, Bosnia-Herzegovina and FYR Macedonia. Clear definition in the first sentence and we already have three good examples of separatist pressure in this first paragraph – already a good Level 2 answer

Closer to home, the Scottish National Party and Plaid Cymru have campaigned for an independent Scotland and Wales respectively. In the case of the former, ☞

the issue is over resources – the SNP feels that the exploitation of North Sea oil and gas has done little to develop the economy of Scotland. Meanwhile, Plaid Cymru wishes to create a bilingual society by promoting the revival of the Welsh language. Both countries feel that they would be able to prosper economically if they were independent and both wish to attain full national status within the European Union. Mec Vannin is a political party operating in the Isle of Man, seeking to revoke the status of the island as a British self-governing Crown dependency and establish a completely independent sovereign state. They campaign strongly against immigration to the island, particularly from England, believing that the population has grown to such an extent that it has increased the burden on the island's infrastructure and environment while eroding the fabric of community life.

A very clear, concise answer where every word counts. An extremely good range of examples with excellent detail in the second paragraph. Top of Level 2: 8 marks.

3 *With reference to examples, outline the issues associated with multicultural societies.* [**10 marks**]

Average answer

A multicultural society is one which contains members from a wide variety of national, linguistic, religious or cultural backgrounds. This can certainly enrich local communities, adding variety to our diet (chicken tikka masala being the most popular 'British' dish), clothing and customs, but it can also lead to problems. The main issues tend to be economic as newcomers to the country struggle to get jobs, especially if they can't speak the local language and may even have to fight against racial prejudice. Local people are very quick to complain that migrants take their jobs by offering their services at a cheaper rate. Friction can also arise if migrants wish to adhere to their own religious calendars and practices. On the other hand, migrants can prove popular with employers because they are prepared to work hard for long hours for relatively low rates of pay. *Top of Level 1 here – a good range of statements and the candidate recognizes some of the positives of a multicultural society. Issues need not be negative!*

There may also be problems with regard to schooling. Again, first-generation migrants may find the language a real barrier, which may in turn mean poor GCSE results which will again put them in a difficult position ☞

when it comes to competing for jobs. Local councils struggle to finance provision for EAL students, and teachers in some inner-city areas find themselves with classes where English is the second language for the majority. *Not really a specific example, but a clear issue well stated which would access Level 2*

There have even been concerns raised about poor health among the migrant community. Immunization has proved a real stumbling block in the past, with migrants feeling suspicious of local practices and unable to understand the importance of vaccination programmes. Low income levels have resulted in poor housing conditions which are themselves associated with a higher level of transmittable disease – increasing levels of TB have been a particular concern.

Again, the final paragraph makes a valid point and does refer to a specific disease (TB). The previous paragraph referred to a specific issue, that of EAL (English as an additional language) provision. Although not case studies, this kind of detail, well expressed, does access Level 2. Overall Level 2: 6 marks.

Strong answer

Multicultural societies are largely formed as a result of migration. The majority of migrants tend to be poor when they first arrive in a country. This fact, coupled with low initial wages, leads to a concentration of ethnic groups in the poorest housing areas of major cities (ghettos). Migrants may find it difficult to obtain employment and to integrate if they do not speak the host country language and may even be subject to discrimination, prejudice and racism. Although migrants tend to be welcome in times of economic growth, during recessions they are often accused of taking local jobs. In addition, the cost of providing for migrants in terms of housing, education and healthcare, may cause resentment and racial intolerance from members of the host population. All things considered, it is not surprising that migrants choose to locate together for reasons of security – having others nearby who speak the same language, have the same customs and religious practices can be an immense comfort. However, it does not encourage integration. No reference to specific examples as yet, but the points made are well explained and quite sophisticated – the answer just accesses Level 2 at this point

Oldham, for example, has long been a town known for attracting migrants, particularly Poles and Ukrainians after the Second World War. However, the South Asian communities which settled from the late 1950s onwards tended to remain much more culturally distinct from the local population, in terms of dress, language and ☞

religion, than previous migrant groups. It may well have been that the differences in colour and ethnicity made it much harder for this group to be assimilated.

The race riots of 2001 followed a long period of inter-racial tensions and attacks in Oldham, with the most violent rioting occuring in the Glodwick area of the town, home to a large community of people of Pakistani heritage. Up to 500 Asian youths were involved, necessitating intervention from 100 police officers. Precise detail again accesses Level 2

Riots also occurred in Bradford slightly later in the same year and can again be traced to the lack of integration with the local community. The district of Manningham, where much of the unrest originated, was 75% South Asian, compared with districts such as Tong and Wibsey which were 93% and 91% white respectively. However, much of the discontent arose within the immigrant communities themselves. Unfortunately, the town council in trying to promote multiculturalism succeeded in creating divisions and tensions within and between different elements of the Asian community, as each fought for a greater allocation of council funding.

A slightly different angle here, recognizing that tensions can emerge within immigrant communities as well as between those communities and the host population – this introduces an element of synopticity which allows the answer to access Level 3: 9 marks.

Essay question

4 *Discuss the economic, social and environmental impacts of one major recent international conflict.*
[**40 marks**]

Strong answer

The Somali civil war is an ongoing conflict which began in 1991, causing destabilization and instability throughout the country. Although primarily a civil war, the conflict has spilled over into other regions and thus can be regarded as 'international'. For example, between 2006 and 2009, the National Defence Force of Ethiopia was involved in the conflict. Somalia's government declared a state of emergency in June 2009, requesting immediate international support and the military intervention of neighbouring East African countries. There has also been heavy involvement from the United Nations. A clear introduction which sets the scene and already displays clear and accurate knowledge of the conflict, which, despite being a 'civil war', has clearly involved several other nations

The first phase of the civil war stemmed from insurrectionist activity against the repressive regime of President Siad Barre, a military dictator. A counter-revolution immediately took place in an attempt to reinstate him as leader of the country. The increasingly violent and chaotic situation has evolved into a humanitarian crisis and to a state of 'anomie'– a complete breakdown of social norms and values. The northern, former British portion of the country actually declared its independence as Somaliland in May 1991, although it has not been recognized as such by any foreign government.

The political situation was and is complex, with rival groups such as the United Somali Congress (USC), the Somali National Movement (SNM) and the Somali Patriotic Movement (SPM) vying for power with armed militia groups, Manifesto, the Somali Democratic Movement (SDM) and the Somali National Alliance (SNA). Divisions among the USC itself led to armed conflict which devastated the capital area of Mogadishu, once known as the 'white pearl of the Indian Ocean'. These two paragraphs are really still introductory, but there is the hint of social and perhaps even environmental impact with this last sentence. Again, a high degree of specific knowledge is evident ☞

The civil war disrupted agriculture and food distribution in southern Somalia, having far-reaching economic and social effects. The resultant famine left about 300 000 dead and caused the UN Security Council to authorise limited peacekeeping operations in 1992; however, as their mandate was limited to self-defence, the force was soon disregarded by the warring clans. In reaction to the continued violence, the United States organized a military coalition with the purpose of creating a secure environment in southern Somalia. This was at least temporarily successful in restoring order and alleviating the famine.

Following the outbreak of the civil war, many of Somalia's residents left the country in search of asylum. At the end of 2009, about 678 000 were under the responsibility of the United Nations High Commissioner for Refugees, constituting the third largest refugee group in the world (only Iraq and Afghanistan can claim more refugees). Due to renewed fighting in the southern half of the country, an estimated 132 000 people left in 2009, and another 300 000 were displaced internally. As is often the case, many of the best educated Somalis left for the Middle East, Europe and North America, seemingly undermining the country's chance of economic recovery. In addition, about 12 000 of the Bantu population, a group traditionally marginalized in Somalia, have sought resettlement in the United States. Inter-related social and economic effects

A further consequence of the collapse of government authority has been the emergence of a significant problem with piracy in the waters off the coast of Somalia. Piracy arose as a response by local fishermen from coastal towns such as Kismayo to illegal fishing by foreign trawlers. However, both local and foreign agencies have been guilty of over-fishing in Somali waters, severely depleting fish stocks. Nor is this the only environmental consequence of the war – lack of stable government has allowed deforestation to proceed at an alarming rate as rival warlords deplete resources for charcoal exports. ☞

Deforestation will have an adverse impact on rainfall availability and the capacity of the soil to hold water. Soil erosion and desertification will be the inevitable result, with a consequent reduction in habitat for animal species and biodiversity. Some of the world's more developed nations have also taken advantage of this political instability and high level of corruption, using the Somali coastline as a dumping ground for hazardous toxic waste materials. Allegedly, both Italian and Swiss firms entered into contracts with Somali warlords and businessmen to dump waste in the country. *Clear environmental effects of the conflict – excellent knowledge and understanding, breadth and depth*

However, it is also possible to argue that the war has had some positive effects – both from a social and an economic standpoint. Although it is difficult to be sure of statistics (and the country did not even appear in the 2009 HDI rankings), the Central Bank of Somalia indicates that the country's GDP per capita is $333, better than that of Tanzania at $280 as well as neighbouring Ethiopia at $100. It is almost as if the civil war has promoted an atmosphere of free enterprise, contrasting sharply with the period beforehand when most services and the industrial sector were government-run. There has certainly been substantial private investment in commercial activities such as food processing and telecommunications, largely financed by the Somali community abroad (the Somali diaspora). Again, according to the Central Bank of Somalia, both imports and exports of goods have surpassed figures recorded at the start of war in 1991. Although there is a trade deficit of about $190 million a year, this appears to be far exceeded by remittances sent by Somalis abroad. This confidence has also spread to foreign multinationals and in 2004, an $8.3 million Coca-Cola bottling plant opened in Mogadishu. Foreign investment has also been forthcoming from General Motors and Dole Fruit. *The positive economic aspects of war – a very unusual approach here and a high level of synopticity. Very clear case study knowledge* ☞

Similarly, although Somalia's public healthcare system was largely destroyed during the early years of the civil war, general living conditions have significantly improved both in absolute terms and relative to other countries in Africa. As with other previously nationalized sectors, informal providers have filled the vacuum and replaced the former government monopoly over healthcare, with access to facilities witnessing a significant increase. Many new healthcare centres, hospitals and pharmacies have been established through home-grown Somali initiatives. In fact, the percentage of the population with access to at least one healthcare facility almost doubled from 28% to 54.8% between 1985 and 2005. Maternal mortality rates also dropped by 30% during this time period, while infant mortality fell from 152 to 115 per 1000 live births. *Positive social factors also clearly identified*

However, the conflict is far from over. In December 2008, Ethiopian soldiers withdrew from Somalia, leaving behind an African Union contingent of several thousand troops to help enforce the fragile authority of the coalition government. Following Ethiopia's withdrawal from Somalia, the southern half of the country rapidly fell into the hands of radical Islamist rebels who established strict Sharia law in areas under their control. The conflict made headline news again recently when a 13–year-old girl was sentenced to death by stoning in Kismayo by the militant group al-Shabab. This would suggest that the social impacts of the conflict remain far-reaching even if the economy is on the road to recovery – and this recovery may itself prove short-lived if it cannot be supported by stable government and effective human rights regimes. The phrase 'no development without security and no security without development' is certainly true for Somalia. *A strong conclusion which brings us back to the problems of warfare, particularly in terms of social impacts. At the same time, the candidate realizes that these factors are interconnected and cleverly uses a phrase from the specification to underline this* ☞

The candidate has addressed the question clearly, providing excellent background detail to the conflict concerned and identifying a range of social, economic and environmental impacts. It is interesting that some of the supposed 'benefits' of the war have also been identified, although one might perhaps have expected more discussion of the social and economic challenges that the country has had to deal with. Case study knowledge is thorough and accurate and the answer is concise, well written and ☞ sophisticated, although the argument could be more direct at times.

Marks:

Knowledge: Level 4 (high)

Understanding: Level 4 (high)

Case studies: Level 4 (high)

Synopticity: Level 4 (high)

Quality of argument: Level 4

Overall mid-Level 4: 38 marks

Plate tectonics and associated hazards

Aftershock	A smaller earthquake that occurs after a previous large earthquake in the same area
Ash	Dust-sized particles of rock produced by the explosive eruption of some volcanoes. This material may be carried in the air for long distances from the volcano which formed it
Asthenosphere	Part of the earth's mantle that lies below the lithosphere, at depths between about 100 and 350 km. The rock here is relatively soft because of its high temperature and relatively low pressure. This enables it to move in a plastic fashion. It is this layer upon which the tectonic plates move
Batholith	A very large mass of igneous intrusive rock (often granite) that forms from cooled magma deep in the earth's crust
Composite volcano	Large, steep-sided, symmetrical cone-shaped volcano formed from alternating layers of lava flows, volcanic ash, cinders, blocks and bombs
Conservative plate margin	A type of plate margin where two tectonic plates are moving past one another with no addition or destruction of plate material
Constructive plate margin	A type of plate margin where new crust is generated as the plates pull away from each other. These are found at mid-oceanic ridges
Continental crust	This is made of old, low density rocks such as granite. It is generally 35–70 km thick and mostly over 1500 million years old
Continental drift	A hypothesis, proposed by Alfred Wegener, that today's continents are the result of the break-up of a single supercontinent. The fragments then drifted to their present positions
Convection current	Currents in the mantle that are the driving force in the movement of the tectonic plates. It is thought that they are initiated by hotspots deep in the mantle
Core	The innermost layers of the earth. The core comprises of two concentric spheres. The inner core is believed to be made of solid iron and the outer core, liquid iron
Crater	A circular depression in the ground caused by volcanic activity
Crust	The outermost layer of the earth
Destructive plate margin	A type of plate margin where crust is destroyed as two plates converge. These are usually associated with island arcs or young fold mountains
Dormant	A description of the state of a volcano between eruptions when it gives out very little gas and lava
Dykes	Steep, sheet-like intrusions, varying in thickness from a few millimetres to tens of metres across. They occupy vertical weaknesses in the rock. They often cut across rock bedding and form low ridges
Epicentre	The point on the earth's surface directly above the focus of an earthquake

Eruption	This is said to occur when a volcano gives off large quantities of lava and gas
Extinct	A description of a volcano that has not erupted for at least 25 000 years
Extrusive activity	Volcanic activity that results from magma reaching the surface
Fault	A fracture in the earth's crust that marks the point where two adjacent masses of rock are moving in different directions
Focus	The point below the surface where an earthquake occurs
Geyser	A type of hot spring that erupts periodically, throwing a column of hot water and steam into the air
Hazard	The potential threat to humans from a naturally occurring process or event
Hot spring	A point where heated groundwater emerges onto the earth's surface
Hotspot	An area deep within the mantle where the temperatures are high enough to initiate convection. They are associated with spreading ridges and isolated chains of volcanic islands found away from plate boundaries
Igneous rock	A rock formed from the cooling of magma
Intrusive activity	Igneous activity that results from the movement of magma within the crust
Island arc	A destructive plate boundary where oceanic crust is subducted beneath oceanic crust
Lahar	A mudflow composed of pyroclastic material and water that flows down from a volcano, usually along a river valley
Lava	Molten rock expelled by a volcano during an eruption
Magma	Molten rock that is found beneath the surface of the earth
Mantle	The layer of the earth between the core and the crust
Moment magnitude scale	A scale used to measure the size of earthquakes in terms of the energy released
Oceanic crust	The type of crust that underlies the ocean basins. It is generally between 5 km and 10 km thick, composed predominantly of basic igneous rock
Ocean trench	A deep depression in the sea floor created at a destructive plate boundary
Palaeomagnetism	A record of the history of the earth's magnetic field, preserved in magnetic minerals in volcanic rocks
Plate tectonics	The theory that states that the earth's crust is made up of several rigid plates moving relative to one another
Plume	A hot column of magma which rises up from deep within the earth
Rift valley	A long, deep valley found in the centre of a spreading ridge. It is formed between parallel faults where a block of the crust has sunk down

Seafloor spreading	The theory that the ocean floor is moving away from the mid-oceanic ridge and across the deep ocean basin, to disappear beneath continents and island arcs
Seismic waves	Shock waves caused by sudden movement along a fault
Shield volcano	A large, low-angled volcano composed of layers of low viscosity basaltic lava
Sill	An igneous intrusion between bedding planes of sedimentary rock layers
Subduction	The process whereby one crustal plate descends below another. This occurs at destructive plate margins
Surface waves	Seismic waves that travel along the surface of the earth. They include Rayleigh waves and Love waves
Tectonic plates	A series of rigid sections of the earth's crust. They float on the upper mantle and move relative to one another
Tephra	Any type of rock fragment that is forcibly ejected from a volcano during an eruption
Tsunami	Sea wave that can be generated by undersea earthquakes, volcanic eruptions and landslides into the sea
Vent	An opening, or rupture, in the earth's surface which allows hot magma, ash and gases to escape from below the surface
Volcanic bombs	Rocks that are more than 5 mm in diameter that are thrown into the air by a volcanic eruption
Volcanic explosive index (VEI)	A scale used to measure the explosiveness of volcanoes
Young fold mountains	Mountains formed at a destructive plate margin or collision zone

Weather and climate and associated hazards

Adiabatic heating/ cooling	The heating/cooling of a gas as a result of pressure and volume changes alone
Air mass	An air mass is an extensive body of air in which there is only gradual horizontal change in temperature and humidity at a given height
Albedo	The reflectivity of a surface
Anthropogenic	Effects that are derived from human activities, as opposed to those occurring in biophysical environments without human influence
Anticyclone	An area of high atmospheric pressure. Anticyclones tend to have a very low pressure gradient and light, variable winds
Atmosphere	The mixture of gases surrounding the earth

Atmospheric pressure	This is the force per unit area exerted against a surface by the weight of air above that surface
Carbon sink	This is a natural or artificial reservoir that accumulates and stores some carbon-containing chemical compounds for an indefinite period
Channelling	Occurs when there are urban 'canyons' which concentrate all airflow in one direction
Climate	The average annual pattern of weather experienced by a place. It is based on records from the last 30 years and describes the seasonal pattern of temperatures and precipitation
Cold front	A boundary between warm and cold air where cold air is advancing on warm air, undercutting it and causing the warm air to rise. Fronts are associated with rainfall
Convectional rainfall	Rainfall resulting from the uplift and subsequent cooling of air that has been heated by contact with a warm land surface
Coriolis force	The effect of the earth's rotation on air flow
Dendrochronology	The use of the annual growth rings of trees to infer past climatic conditions. Counting the rings can also give us a date
Depression	An area of low atmospheric pressure with a roughly circular pattern of isobars that occurs in temperate latitudes
Ferrel cell	The atmospheric convection cell between the subtropical high pressure zone and the temperate low pressure zone
Geographical model	A model constructed to explain overall patterns rather than localized variations, e.g. the tri-cellular model
Global warming	A term used to describe the recent, rapid rise in global temperatures
Greenhouse gas	An atmospheric gas that acts as a filter, allowing incoming shortwave ultraviolet and light radiation through the atmosphere but stopping long wave infrared radiation from leaving, e.g. CO_2
Hadley cell	The atmospheric convection cell between the equatorial low pressure zone and the subtropical high pressure zone
Heat budget	This is the balance between the incoming solar radiation (insolation) and outgoing radiation from the planet
Insolation	A measure of solar radiation energy received on a given surface area in a given time
Inter tropical convergence zone (ITCZ)	The low pressure equatorial region where there is rising air. It is located where the NE trade winds meet (converge with) the SE trade winds
Jet stream	A narrow belt of fast-moving air near the top of the tropopause
Lapse rate	The rate at which temperature decreases with height

Latent heat	The amount of energy released or absorbed by a substance during a change of physical state that occurs without changing its temperature
Latitude	The angular distance of a place north or south of the equator
Maritime	The influence exerted by seas and oceans. It tends to have a moderating effect on climate
North Atlantic Drift	The northwest extension of the Gulf Stream. It brings slow-moving warm water to NW European shores
Orographic rainfall	Rainfall resulting from the uplift and subsequent cooling of air over high ground
Particulate pollution	A term used to describe particles of 10 micrometres or less that are the result of human activity, particularly industrial processes and vehicle exhausts. They can include cement dust, tobacco smoke, ash and coal dust
Photochemical smog	A form of air pollution caused by a photochemical reaction between the exhaust gases of cars and sunshine
Polar cell	The atmospheric convection cell between the polar high pressure zone and the temperate low pressure zone
Prevailing wind	The most common wind direction for a location
Rossby waves	A series of waves that occur in the upper westerly winds which blow in the higher parts of the atmosphere. They occur in both hemispheres
Sahel	This is the area between the Sahara desert in the north and the savanna in the south. It stretches across the north of the African continent between the Atlantic Ocean and the Red Sea
Savanna	The name given to a climate which can be found in tropical sub-Saharan Africa, the Brazilian Plateau and northern Australia
Subtropical high pressure	The area of high pressure found between 25° and 35° N and S of the equator
Temperate	This describes temperatures or climates with few extremes. It can also describe the latitudes between the tropics and the polar circles
Trade winds	This is a reference to the pattern of prevailing easterly surface winds found in the tropics
Tri-cellular model	A model that explains some of the main aspects of atmospheric circulation. It divides each hemisphere into three large convection cells
Tropical revolving storm	A generic term that refers to intense low pressure weather systems that originate over warm tropical oceans and migrate westward and poleward. They have a variety of names, e.g. hurricane
Tropopause	The boundary between the troposphere and the stratosphere
Troposphere	The lowest layer of the atmosphere. It varies in thickness between 8 km and 12 km. It contains 75% of the atmospheric gases

Urban heat island	The zone around and above an urban area which has higher temperatures than the surrounding rural area
Venturi effect	The squeezing of moving air through a narrow gap (between buildings) that increases the velocity of the wind
Warm front	A boundary between warm and cold air where warm air is advancing on cold air. The less dense warm air rises over the denser cold air. Fronts are associated with rainfall
Weather	The hour-by-hour state of the atmosphere in any one place

Ecosystems: change and challenge

Abiotic factors	Influences on an ecosystem from non-living things in the environment, e.g. water, light, warmth, humidity, rocks and soils
Autotroph	Any organism that can synthesize its food from inorganic substances, using heat or light as a source of energy
Biodiversity	The variation of life forms within a given ecosystem, biome, or for the entire earth. Biodiversity is often used as a measure of the health of biological systems
Biomass	The mass of living biological organisms in a given area or ecosystem at a given time. Biomass can refer to species biomass, which is the mass of one or more species, or to community biomass, which is the mass of all species in the community
Biome	A major habitat category, based on distinct plant assemblages which depend on particular temperature and rainfall patterns, e.g. tundra, temperate forest and rainforest etc.
Biotic factors	Influences on an ecosystem from living things, e.g. plants, animals, fungi and bacteria
Carnivore	An organism that derives its energy and nutrient requirements from a diet consisting mainly or exclusively of animal tissue, whether through predation or scavenging
Climatic climax	A biological community of plants and animals which, through the process of ecological succession, has reached a state of dynamic equilibrium with its climate and soils
Colonization	The process in ecology by which a species spreads into new areas
Conservation	Preservation of the natural environment
Consumers	These are organisms that feed on other organisms below them in the food chain
Deciduous	Plants or trees that lose their leaves for part of the year, usually winter, to reduce transpiration and conserve water
Deforestation	The deliberate clearing of forest by either cutting or burning

Ecological footprint	A measure of how much biologically productive land and water area an individual, population or activity requires to produce all the resources it consumes and to absorb its waste using current technology and resource management practices
Ecology	The study of communities of living organisms and the relationships among the members of those communities, and between them and their surroundings
Ecosystem	A system in which organisms interact with each other and the environment
Epiphyte	A plant that derives moisture and nutrients from the air and rain; usually grows on another plant but not parasitic on it
Evapotranspiration	The combined output of water from an area through the processes of evaporation and transpiration
Extractivist	This describes an economic activity or area used by traditional communities based on small-scale extractive activities (such as harvesting rubber and Brazil nuts) and subsistence agriculture
Food chain	An arrangement of the organisms of an ecological community according to the order in which they eat each other, with each organism using the next lower organism in the food chain as a source of energy
Food web	A scheme of feeding relationships, resembling a web, which unite the member species of a biological community
Gersmehl diagram	A diagram to illustrate the mineral nutrient cycle
Habitat	An ecological or environmental area that is inhabited by a particular species of animal, plant or other type of organism
Herbivore	Any organism that eats only plant material
Inter-tropical convergence zone (ITCZ)	The zone of low atmospheric pressure located between the tropics where the NE and SE trade winds meet. It is an area of high temperatures and convectional rainfall
Lithosere	A plant succession that begins life on a newly exposed rock surface, such as one left bare as a result of glacial retreat
Litter	Dead organic material that has not been fully decomposed which gathers on the surface or in upper levels of soil
Microhabitat	A small, narrowly defined and specialized habitat occupied by a species
Niche	The position of a species within an ecosystem, describing both the range of conditions necessary for persistence of the species, and its ecological role in the ecosystem
Nutrient cycle	The movement of nutrients in the ecosystem between the three major stores of the soil, biomass and litter
Photosynthesis	The process in green plants and certain other organisms by which carbohydrates are synthesized from carbon dioxide and water using light as an energy source

Plagioclimax	The plant community that exists when human interference prevents the climatic climax vegetation being reached
Plant succession	The changes in the composition of a community of plants over time. It refers to the sequence of communities which replace one another in a given area
Prisere (primary succession)	The development of a plant community by the gradual colonization of a lifeless abiotic surface
Producers	Organisms that can manufacture food from inorganic raw materials. They are also called autotrophs
Secondary succession	Succession taking place on an area that had formerly been vegetated but has undergone loss of that vegetation
Sere	The entire sequence of stages in a plant succession. Different seres are named after the starting point of the succession, e.g. lithosere, hydrosere, psammosere and halosere
Seral stage	An individual stage within a sere, e.g. colonization or stabilization
Soil	The top layer of the earth's surface, containing unconsolidated rock and mineral particles mixed with organic material, air and water. It provides the foundation for all plant life
Species diversity	See biodiversity
Tertiary consumer	A high-level consumer, which is usually the top predator in an ecosystem and/or food chain
Transpiration	The loss of water from parts of plants especially leaves, but also stems, flowers and roots
Trophic level	An organism's position in the food chain. Level 1 is formed of autotrophs which produce their own food. Level 2 organisms consume level 1 and level 3 consume level 2 etc.

World cities

City Challenge partnerships	These were based on a system of competitive bidding by local authorities who had to develop imaginative plans involving the private sector and the local community to gain funding
Counter-urbanization	This is the migration of people from major urban areas to smaller towns, villages or rural areas – often 'leap-frogging' the green belt
Energy recovery	The use of methods such as incineration and composting to turn unwanted waste into useful energy
Gentrification	This is the process by which older, often rundown housing areas (usually close to the city centre) become desirable once again and are physically and socially upgraded
Infrastructure	The basic facilities, services and installations needed for the functioning of a community or society
Megacity	A city with more than 10 million inhabitants
Millionaire city	A city with more than 1 million inhabitants
Out-of-town retailing	Born from the desire for one-stop shopping. Large retail centres offering a multiplicity of services have developed on the outskirts of large towns and cities and often have their own motorway exits for ease of access
Prestige project developments	Also known as 'flagship projects' as they involve the creation of innovative and eye-catching developments which aim to lead the way in regenerating areas
Property-led regeneration	A form of redevelopment largely associated with urban development corporations (UDCs), with the intention that private investment would be four to five times greater than the public money initially invested
Re-urbanization	This is the movement of people back into urban areas, particularly the inner city or even the CBD itself
Suburbanization	This is the process of population movement (and increasingly industry and retail) from the central areas of cities to the outskirts (the suburbs), often engulfing surrounding villages/rural areas
Sustainability	Sustainability is development that meets the needs of the present without compromising the ability of future generations to do the same. It can be categorized into economic, environmental and social sustainability
Urban development corporation (UDC)	UDCs were set up in the 1980s and 1990s to take responsibility for the physical, economic and social regeneration of selected inner-city areas with large amounts of derelict and vacant land
Urbanization	This is the growth in the proportion of a country's population that lives in towns/cities as opposed to rural areas
Waste management	The clearance of the unwanted by-products of human activity. This is a key sustainability issue as much waste presents health dangers to people and other organisms, as well as problems of using up space

World city	A city which has global influence as a major centre for finance, trade, politics and culture

Development and globalization

Asian Tigers	The Asian Tigers are a group of countries with highly developed economies; they were the original NICs. They are Hong Kong, Singapore, South Korea and Taiwan
Bilateral aid	Bilateral aid is aid that has been directly given from one country to another country in need
Bottom-up schemes	Aid schemes that are focused at the community level and address local needs
Brandt line	The imaginary line separating the 'rich north' from the 'poor south'
BRIC countries	Brazil, Russia, India and China – countries deemed to be at a similar stage of newly advanced economic development
Common markets	Similar to customs unions, but also allowing the free movement of labour and capital
Customs unions	Members impose a tariff on imports from outside the group
Development continuum	A recently introduced term to describe the difference between developed and undeveloped countries. This term places greater emphasis on recognizing the increasing gap between the rich and poor countries
Development gap	This refers to the difference in wealth (and associated development indicators such as life expectancy) between developed and developing countries
Economic unions	Similar to common markets, but these groups also agree to adopt common policies on certain issues such as pollution. The European Union (EU) is an example
Ecotourism	Another name for **sustainable tourism**
Emerging markets	Nations that are experiencing rapid growth and industrialization. This term refers to countries such as China
EU	The European Union
Eurozone	The 16 members of the EU that have adopted a common currency, the Euro
FDI	Foreign direct investment
Free trade areas	Countries within free trade areas are able to trade freely without being subject to quotas or tariffs, but members still impose a tariff on goods coming from outside the group
Global products	Products that are sold all over the world in the same form, e.g. Coca-Cola
Globalization	A process whereby regional economies, societies and cultures become integrated into a global network through increasing ease of communication, transport and trade

Gross Domestic Product (GDP)	The total value of goods and services produced by a country in a year – usually expressed per capita
Gross National Income (GNI)	The total value of goods and services produced within a country, together with its net income received from other countries
Gross National Product (GNP)	As GDP *plus* all net income earned by that country and its population from overseas sources
HIPCs	Heavily indebted poor countries
Human Development Index (HDI)	A statistic used to rank countries by their level of 'human development', with countries separated into developed, developing and underdeveloped countries. This is based on the GDP per capita, life expectancy and education of the countries concerned
IMF	International Monetary Fund
'Just in time' production	Production that aims to increase efficiency and reduce stockpiling in warehouses. It relies on communication and signals between parts of the organization to indicate when certain parts or whole products need to be produced
LDCs	Least developed countries
Long-term aid	The aim of this sort of aid is to alleviate poverty in the developing country. The aid is often used to help promote development within a country, be that social or economic development
MDGs	Millennium Development Goals
MDRI	Multilateral Debt Relief Initiative
Multilateral aid	Multilateral aid is aid that has been given from a country to another country via an international organization such as the World Bank
NGO	Non-governmental organization
NICs	Newly industrialized countries
North–south divide	The imaginary line separating the 'rich north' from the 'poor south' – often known as the **Brandt line**
ODA	Official development assistance
Offshore outsourcing	This is the practice of hiring an external organization to perform certain business functions in a country other than the one where the products or services were originally developed. Also known as 'offshoring'
PRSP	Poverty Reduction Strategy Papers
Quality of life	A term used to describe the general wellbeing of individuals and societies
RICs	Recently industrialized countries
SEZs	Special economic zones

Short-term aid	When a country gives assistance to another country that is in immediate need of aid, e.g. following a natural disaster. Its aim is to alleviate suffering in the short term
Sustainable development	This is development which meets the needs of the present without compromising the needs of future generations
Sustainable tourism	Sector involved in promoting tourism in many areas of the world, but in such a way that it has a minimal effect on the environment
Tied aid	Aid that must in part be used to buy products from the country that donated the aid
TNCs	Transnational corporations
Top-down aid	Often large-scale projects that are directed from the 'top' of governments or international aid organizations
UN	The United Nations
WTO	World Trade Organization

Contemporary conflicts and challenges

Appropriate/intermediate technology	Technology which is affordable, available locally and utilizes local skills and materials
Autonomy	Self-government
Civil disobedience	The active refusal of an individual or group of individuals to obey the laws of the government or occupying international organization
Civil war	A war fought between different groups from and within the same country
Conflict	In political terms, 'conflict' refers to an ongoing state of hostility between two or more groups of people; in a geographical context, conflict is often about the best way to utilize resources
Culture	This refers to the customary beliefs, social norms and traits of a racial, religious or social group
Ethnic segregation	The clustering together of people with similar ethnic or cultural characteristics into separate urban residential areas
Ethnicity	This is the grouping of people according to their ethnic origins or characteristics
Gross Domestic Product (GDP)	The total value of goods and services produced by a country in a year – usually expressed per capita
Gross National Product (GNP)	As above *plus* all net income earned by that country and its population from overseas sources

Human Development Index (HDI)	A statistic used to rank countries by their level of 'human development' with countries separated into developed, developing and underdeveloped countries. This is based on the GDP per capita, life expectancy and education of the countries
Identity	This refers to a sense of belonging to a particular group or geographical area
Ideology	A set of ideas that constitute an individual's or a group's goals, expectations and related actions
Insurrection	The act of open revolt against civil authority or local government. Those that take part in insurrection are called insurgents
International migration	The movement of people across national frontiers involving a permanent change of residence
International poverty line	This is the minimum level of income considered necessary to achieve an adequate standard of living in a given country
Market processes	Processes whereby the ability to pay the asking price for a local resource outweighs local and national concerns over how said resource should be used
Millennium Development Goals	Global action plan instigated by the UN to achieve eight anti-poverty goals by a 2015 target date
Multicultural society	A social grouping which contains members from a wide variety of national, linguistic, religious or cultural backgrounds
Physical Quality of Life Index (PQLI)	This index makes an attempt to quantify the quality of life for people in a country. It is based on three statistics: literacy, infant mortality and life expectancy
Planning processes	Processes whereby those wishing to utilize a local resource must take into consideration the views of others, such as the local authority
Poverty	The condition of not having the means to afford basic human needs such as clean water, nutrition, healthcare, education, clothing and shelter
Purchasing power parity	A calculation whereby the purchasing power of different currencies is equalized for a given basket of goods
Refugee	Someone who is no longer in their home country due to fear of persecution because of their race, religion, nationality, political or social group. Refugees may also have fled their homes due to civil war, ethnic, tribal or religious violence or environmental disasters
Separatist pressure	The pressure by a group of people within one or more countries to gain greater autonomy
Territory	A geographical area belonging to, or under the jurisdiction of, a governmental authority
Terrorism	The use of violence and/or threats to intimidate or coerce others
World Bank	A source of financial and technical assistance to more than 100 developing countries around the world